DEMYSTIFYING
CLIMATE
CHANGE
TERMS & CONCEPTS

A COMPENDIUM

Dr. Sonia Dipti M.V
Dr. Pooja Dipti M.V
Prof. Velmurugan P.S

DISCLAIMER

FOREWORD

ExNoRa® International Foundation

New No.40, Vijayaragava Road, Rams Flats. Flat No.1, Second Floor, T.Nagar, Chennai - 600 017,
Phone : 044-42121673 E-mail : exnora@exnorainternational.org.in Website : www.exnorainternational.org.in

In an era when climate change dictates the trajectory of our planet's future Climate Change Demystifying Terms & Concepts emerges as an indispensable guide. Authored by Prof. Velmurugan P. S. and Dr. Sonia Dipti Velmurugan, this book masterfully simplifies the intricate science of climate change, ensuring accessibility for readers across diverse backgrounds.

The book thoughtfully navigates key topics, from foundational ideas like greenhouse gases and carbon footprints to critical subjects such as climate justice and adaptive strategies. Its content, rooted in comprehensive research, offers a panoramic understanding of the challenges and potential solutions in addressing the climate crisis.

What sets this book apart is its actionable focus, perfectly aligning with ExNoRa International Foundation's philosophy of catalysing change through deeds, not just discussions. ExNoRa, renowned for its community-driven environmental initiatives, underscores the importance of turning awareness into impactful action—a vision this book wholeheartedly embraces. While its scientific rigour is commendable, the book's true strength lies in its ability to transform understanding into action. It emphasizes not just the urgency of the crisis but also the role each of us plays in building a sustainable and equitable world. It highlights the disproportionate burdens of climate change while celebrating the resilience and ingenuity of vulnerable communities.

This is far more than a theoretical exploration—it is a rallying cry for collective action. The book inspires readers to move beyond passive learning and actively contribute to mitigating climate change through informed decisions and community efforts.

Let this book serve as a beacon, urging us all to act with clarity, compassion, and determination as we confront one of humanity's most pressing challenges. Together, we can transform knowledge into impactful, lasting change.

S. Senthur Pari,
President - ExNoRa International Foundation

FOREWORD

The authors of this book, Dr. Sonia Dipti M.V., Dr. Pooja Dipti M.V and Dr.Velmurugan.P.S., are eminent scholars in their respective fields. Dr. Velumurugan, as a professor, has proven track record of knowledge and literary work in multidisciplinary platforms as a professor at Department of Commerce, Cantal University of Tamil Nadu, Thiruvarur, Tamil Nadu. His vast exposure out of national and international research project made him to be an expert in the field of climate change and environment. Dr. Sonia is an additional highly valuable resource with a passion of writing besides professional task of dentist, researcher, and informaticist based in Philadelphia, Pennsylvania, USA. Having common interest and passion toward the climate and environment, both the authors have conceived noteworthy writing project in the form of book titled " Climate Change : Demystifying Terms & Concepts"

In fact, the climate change and environmental destruction are major threats to the future of our planet. The sustainability of life on Earth is under increasing threat due to human induced climate change. This perilous change in the Earth's climate is caused by increases in carbon dioxide and other greenhouse gases in the atmosphere, primarily due to emissions associated with burning fossil fuels. Over the next two to three decades, the effects of climate change, such as heatwaves, wildfires, droughts, storms, and floods, are expected to worsen, posing greater risks to human health and global stability.

In this context, as a President of India Vetiver Foundation (IVF), I wish to mention that The role of vetiver in building climate resilience is a enormous. We may think that the vetiver grass looks like any other wild grass at a first glance, studies have shown that vetiver is a cost-effective and eco-friendly slope stabilisation mechanism which can help mitigate soil erosion, flooding and landslides. In the 1993, the National Research Council (NRC) has published a scientific audit of the safety and effectiveness of Vetiver Grass for erosion control. Dr. Norman Borlaug (Noble Laureate) chaired the Panel which concluded that 'accumulated experiences… add up to a compelling case that vetiver is one practical, and probably powerful, solution to soil erosion for many locations throughout the warmer parts of the word'.

In the light of above, the attempt of Dr.Velmurugan.P.S. and Dr. Sonia Dipti Meena Velmurugan in writing this book is highly appreciated. The book has covered all relevant terminalizes of climate change and environment. The present of the book contains with relevant pictures, graphs, diagrams and flow charts. These are very simple and essay to understand by the readers. This book compiles all important available information related toimplementation of mitigation and adaptation strategies, pollution and environmental degradation and summarized evidence of climate change in Earth's spheres, discuss emission pathways and drivers of climate change, and analyse the impact of climate change on environmental and human health.

Happy Reading…

(Dr. C.K. Ashok Kumar)
President
India Vetiver Foundation (IVF)

PREFACE

The Industrial Revolution transformed humanity, supported by energy sources, and changed economies that had been based on agriculture and handicrafts into economies based on large-scale industry, manufacturing, and the factory system. Human advancements in technology, healthcare, and governance, backed by research and development and innovation, have improved the quality of life and increased Human life expectancy. However, the use of greenhouse gas-emitting energy sources and the production of newer materials accelerated the current rapid warming trend, which is largely attributed to human activities since that period. Lawmakers from around the world understand that the critical phenomenon of climate change is significantly different from the variations that have occurred throughout Earth's history, based on findings produced by global scientists and researchers.

There have been several climate change conferences, meetings, and pacts signed ever since these findings started to alarm humanity. Nations have promised to work towards reducing factors contributing to this phenomenon, such as reducing the usage of fossil fuels, and increasing factors that will mitigate it, like adopting renewable and sustainable energy sources. Climate change is one of the most critical challenges of our time, affecting every facet of life on Earth. This book provides an in-depth exploration of the concepts relating to climate change focusing on the issue, addressing its causes, far-reaching consequences, and potential solutions. It included concepts and terms on both natural processes and human activities driving climate change, alongside the environmental, social, and economic risks posed by inaction.

The purpose of this book is to make the complexities of climate change accessible to a broad audience, bridging the gap between scientific knowledge and public understanding at a very basic level by compiling the concepts from existing published works. For instance, the compendium underscores the intricate relationships between human activities—such as transportation, agriculture, and energy use—and their impact on the planet's delicate climate systems. The book also sheds light on issues of climate justice, revealing how marginalized populations often face the worst impacts of climate change despite contributing the least to the problem.

This book compiles all important available information from the web and data, publications, news from around the world, and events about this topic, presenting it to readers as a one-stop publication for those interested in learning about climate change in depth. Intended for students, policymakers, researchers, and concerned citizens, this compendium offers a comprehensive look at the scientific, historical, and ethical aspects of climate change. By fostering a deeper understanding of these dimensions, it encourages informed decisions and collective action. Whether you are new to the subject or an expert seeking fresh perspectives, this book provides insights to help guide efforts toward a sustainable and equitable future.

Through a combination of practical solutions, real-world examples, and analysis of global trends, this book empowers readers to take an active role in addressing climate change. It emphasizes the urgency of collaboration and innovation in building resilient communities, protecting ecosystems, and reducing future risks. By enhancing understanding and inspiring action, the concepts and terms compiled in the forthcoming pages aspires to contribute meaningfully to the global response to one of the greatest challenges humanity has ever faced.

-THE EDITORS

Table of Contents

INTRODUCTION

The Industrial Revolution transformed humanity, supported by energy sources, and changed economies that had been based on agriculture and handicrafts into economies based on large-scale industry, manufacturing, and the factory system. Human advancements in technology, healthcare, and governance, backed by research and development and innovation, have improved the quality of life and increased Human life expectancy. However, the use of greenhouse gas-emitting energy sources and the production of newer materials accelerated the current rapid warming trend, which is largely attributed to human activities since that period. Lawmakers from around the world understand that the critical phenomenon of climate change is significantly different from the variations that have occurred throughout Earth's history, based on findings produced by global scientists and researchers. There have been several climate change conferences, meetings, and pacts signed ever since these findings started to alarm humanity. Nations have promised to work towards reducing factors contributing to this phenomenon, such as reducing the usage of fossil fuels, and increasing factors that will mitigate it, like adopting renewable and sustainable energy sources. Climate change is one of the most critical challenges of our time, affecting every facet of life on Earth. This book provides an in-depth exploration of the concepts relating to climate change focusing on the issue, addressing its causes, far-reaching consequences, and potential solutions. It included concepts and terms on both natural processes and human activities driving climate change, alongside the environmental, social, and economic risks posed by inaction.

The purpose of this book is to make the complexities of climate change accessible to a broad audience, bridging the gap between scientific knowledge and public understanding at a very basic level by compiling the concepts from existing published works. For instance, the compendium underscores the intricate relationships between human activities—such as transportation, agriculture, and energy use—and their impact on the planet's delicate climate systems. The book also sheds light on issues of climate justice, revealing how marginalized populations often face the worst impacts of climate change despite contributing the least to the problem.

This book compiles all important available information from the web and data, publications, news from around the world, and events about this topic, presenting it to readers as a one-stop publication for those interested in learning about climate change in depth. Intended for students, policymakers, researchers, and concerned citizens, this compendium offers a comprehensive look at the scientific, historical, and ethical aspects of climate change. By fostering a deeper understanding of these dimensions, it encourages informed decisions and collective action. Whether you are new to the subject or an expert seeking fresh perspectives, this book provides insights to help guide efforts toward a sustainable and equitable future.

Through a combination of practical solutions, real-world examples, and analysis of global trends, this book empowers readers to take an active role in addressing climate change. It emphasizes the urgency of collaboration and innovation in building resilient communities, protecting ecosystems, and reducing future risks. By enhancing understanding and inspiring action, the concepts and terms compiled in the forthcoming pages aspires to contribute meaningfully to the global response to one of the greatest challenges humanity has ever faced.

GLOBAL WARMING

Global warming is the long-term warming of the planet's overall temperature. Though this warming trend has been going on for a long time, its pace has significantly increased in the last hundred years due to the burning of fossil fuels. As the human population has increased, so has the volume of fossil fuels burned. Fossil fuels include coal, oil, and natural gas, and burning them causes what is known as the "greenhouse effect" in Earth's atmosphere. The greenhouse effect is when the sun's rays penetrate the atmosphere, but when that heat is reflected off the surface cannot escape back into space. Gases produced by the burning of fossil fuels prevent the heat from leaving the atmosphere.

These greenhouse gasses are carbon dioxide, chlorofluorocarbons, water vapor, methane, and nitrous oxide. The excess heat in the atmosphere has caused the average global temperature to rise overtime, otherwise known as global warming. Global warming has presented another issue called climate change. Sometimes these phrases are used interchangeably,

however, they are different. Climate change refers to changes in weather patterns and growing seasons around the world. It also refers to sea level rise caused by the expansion of warmer seas and melting ice sheets and glaciers. Global warming causes climate change, which poses a serious threat to life on Earth in the forms of widespread flooding and extreme weather.

Scientists continue to study global warming and its impact on Earth. The extreme droughts, wildfires, floods, tropical storms, and other disasters that we refer to collectively as climatechange. These effects are felt by all people in one way or another but are experienced most acutely by the underprivileged, the economically marginalized, and people of colour, for whom climate change is often a key driver of poverty, displacement, hunger, and social unrest. Global warming occurs when carbon dioxide (CO_2) and other air pollutants collect in the atmosphere and absorb sunlight and solar radiation that have bounced off the earth's surface. Normally this radiation would escape into space, but these pollutants, which can last for years to centuries in the atmosphere, trap the heat and cause the planet to get hotter. These heat-trapping pollutants—specifically carbon dioxide, methane, nitrous oxide, water vapor, and synthetic fluorinated gases—are known as greenhouse gases, and their impact is called the greenhouse effect. Though natural cycles and fluctuations have caused the earth's climate to change several times over the last 800,000 years, our current era of global warming is directly attributable to human activity—specifically to our burning of fossil fuels such as coal, oil, gasoline, and natural gas, whichresults in the greenhouse effect. In the United States, the largest source of greenhouse gases is transportation (29 percent), followed closely by electricity production (28 percent) and industrial activity (22 percent). Learn about the natural and human causes of climate change.Curbing dangerous climate change requires very deep cuts in emissions, as well as the use of alternatives to fossil fuels worldwide. The good news is that countries around the globe have formally committed—as part of the 2015 Paris Climate Agreement—to lower their emissions by setting new standards and crafting new policies to meet or even exceed those standards. The not-so-good news is that we're not working fast enough. To avoid the worst impacts of climate change, scientists tell us that we need to reduce global carbon emissions by as much as 40 percent by 2030. For that to happen, the global community must take immediate, concrete steps: to decarbonize electricity generation by equitably transitioning from fossil fuel–based production to renewable energy sources like wind and solar; to electrify our cars and trucks; and to maximize energy efficiency in our buildings,

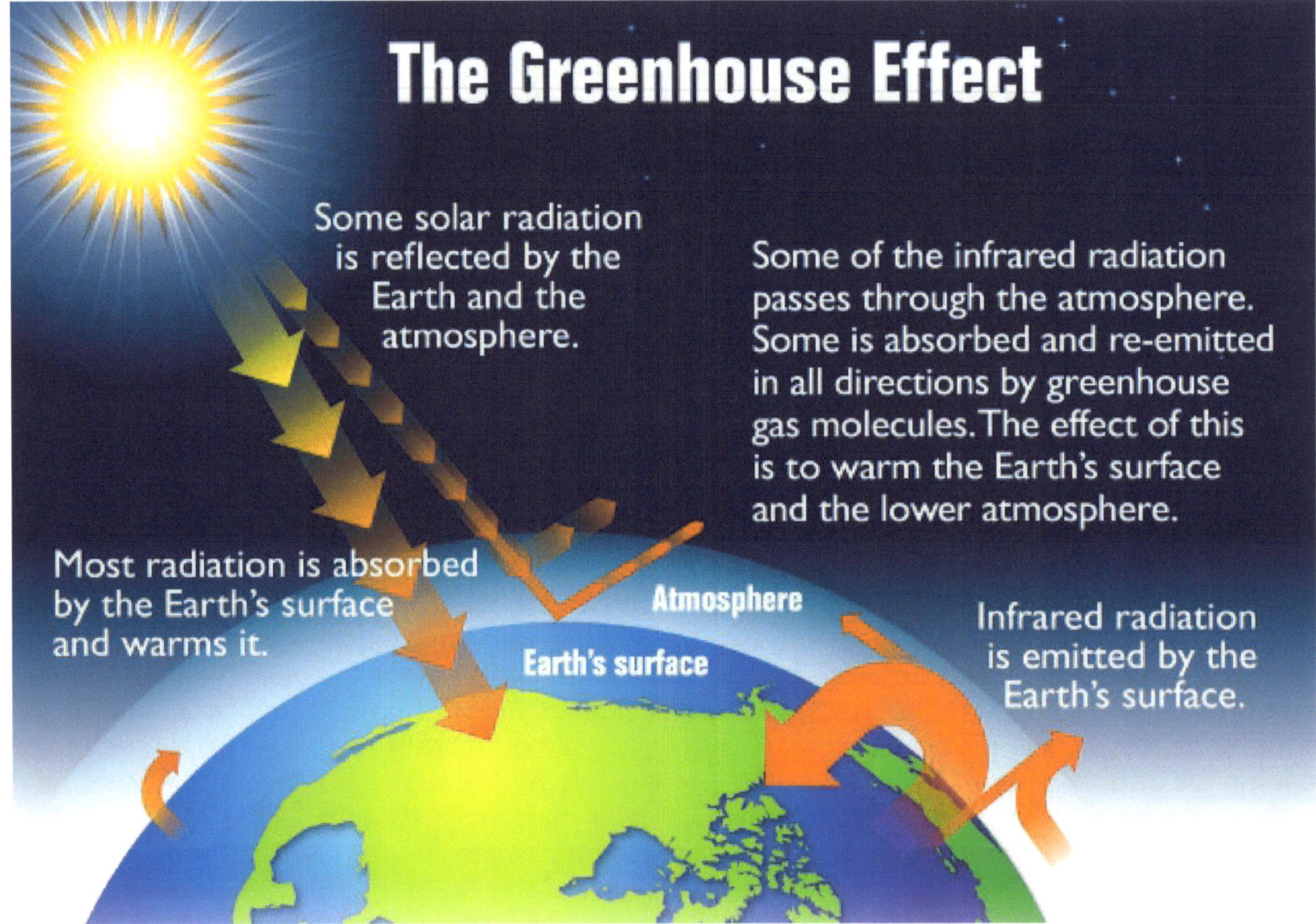

appliances, and industries. The global mean temperature in 2024 is on track to outstrip the temperature even of 2023, the current warmest year. For 16 consecutive months (June 2023 to September 2024), the global mean temperature likely exceeded anything recorded before, and often by a wide margin, according to WMO's consolidated analysis of the datasets. One or more individual years exceeding 1.5°C does not necessarily mean that "pursuing efforts to limit the temperature increase to 1.5°C above pre-industrial levels" as stated in the Paris Agreement is out of reach. The exceedance of warming levels referred to in the Paris Agreement should be understood as an exceedance over an extended period, typically decades or longer, although the Agreement itself does not provide a specific definition. As global warming continues, there is an urgent and unavoidable need for careful tracking, monitoring and communication with regard to where the warming is relative to the long-term temperature goal of the Paris Agreement, to help policymakers in their deliberations. To support this, WMO has established an international team of experts, and the initial indication is that long-term global warming is currently likely to be about 1.3°C compared to the 1850-1900 baseline.

WHAT IS THE GREENHOUSE EFFECT?

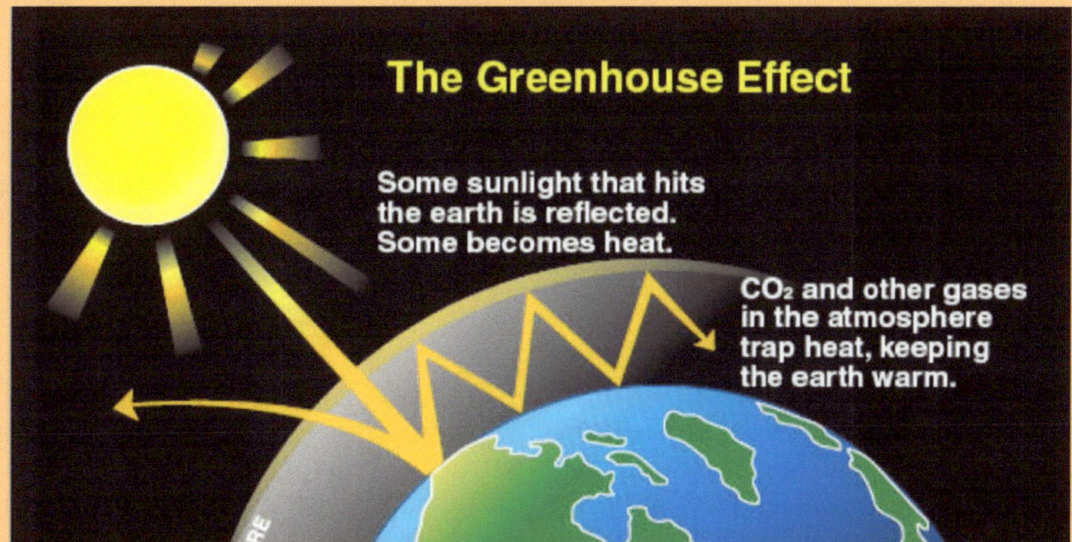

The main driver of climate change is the greenhouse effect. Greenhouse gases in the Earth's atmosphere act a bit like the glass in a greenhouse, trapping the sun's heat and stopping it from leaking back into space and causing global warming.

Source: https://karmawallet.io/blog/2024/04/impact-insider-global-warming/ & https://hips.hearstapps.com/hmg-prod/images/concept-illustration-global-warming-around-the-royalty-free-image-1707420637.jpg
Reference link: https://education.nationalgeographic.org/resource/global-warming/   https://nrdc.org/stories/global-warming-101#causes, https://climateconnection.org.in/resources/climate-change-green-house-effect,
https://www.epa.gov/climatechange-science/basics-climate-change

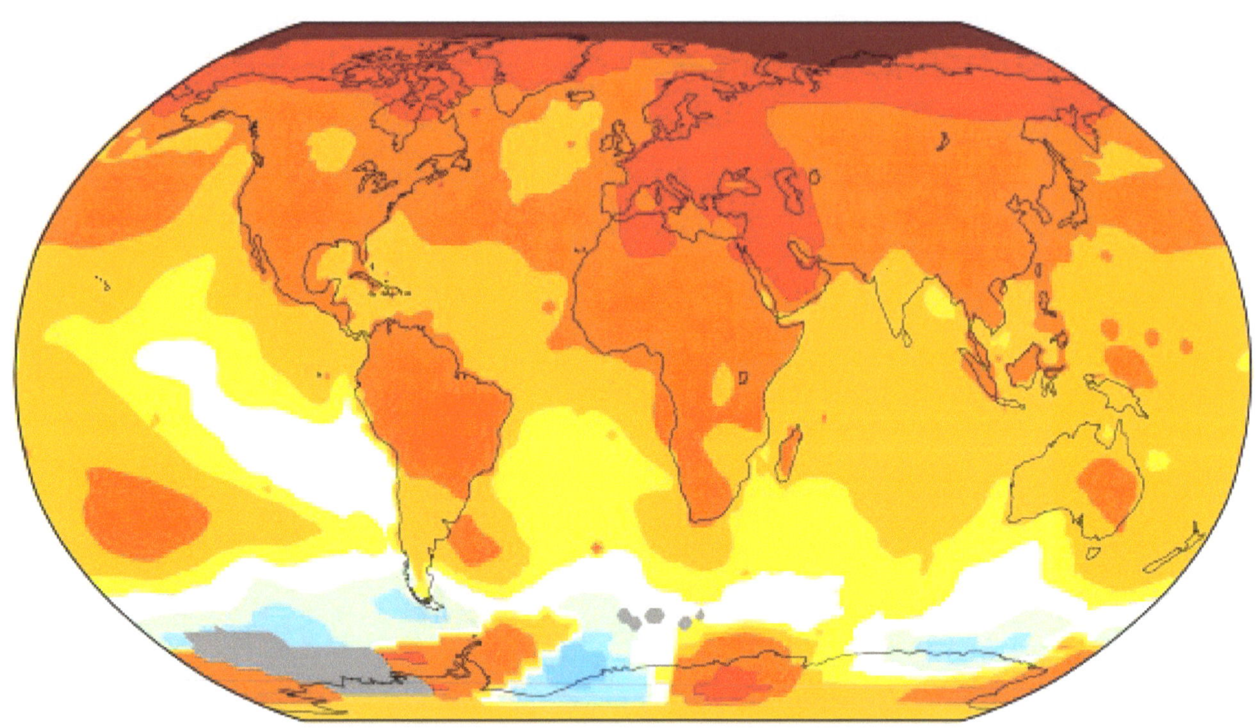

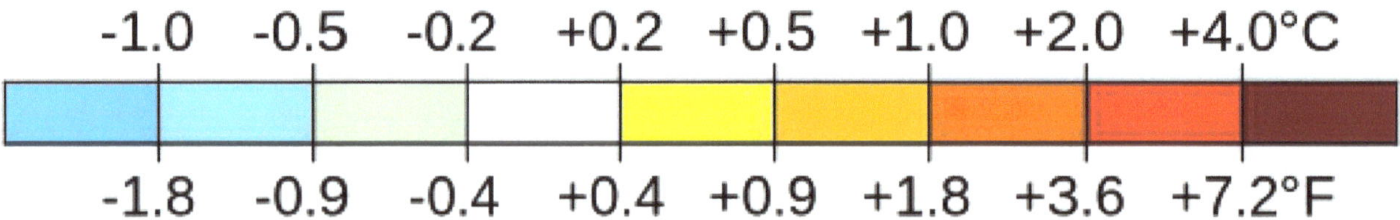

Climate change refers to long-term shifts in temperatures and weather patterns. Such shifts can be natural, due to changes in the sun's activity or large volcanic eruptions. But since the 1800s, human activities have been the main driver of climate change, primarily due to the burning of fossil fuels like coal, oil and gas. Burning fossil fuels generates greenhouse gas emissions that act like a blanket wrapped around the Earth, trapping the sun's heat and raising temperatures. The main greenhouse gases that are causing climate change include carbon dioxide and methane. These come from using gasoline for driving a car or coal for heating a building. For example, Clearing land and cutting down forests can also release carbon dioxide. Agriculture, oil and gas operations are major Sources of methane emissions. Energy, industry, transport, buildings, agriculture, oil and gas operations are major Sources of methane emissions. Energy, industry, transport, buildings, agriculture and land use are among the main sectors causing greenhouse gases. Changes in surface air temperature over the past 50 years. The Arctic has warmed the most, and temperatures on land have generally increased more than sea surface.

CAUSES OF CLIMATE CHANGE

NATURAL CAUSES

S ome amount of climate change can be attributed to natural phenomena. Over the course of Earth's existence, volcanic eruptions, fluctuations in solar radiation, tectonic shifts, and even small changes in our orbit have all had observable effects on planetary warming and cooling patterns. But climate records are able to sho=w that today's global warming particularly what has occurred since the start of the industrial revolution is happening much, much faster than ever before. According to NASA, these natural causes are still in play today, but their influence is too small or they occur too slowly to explain the rapid warming seen in recent decades." And the records refute the misinformation that natural causes are the main culprits behind climate change, as some in the fossil fuel industry and conservative think tanks would like us to believe.

HUMAN-DRIVEN CAUSES

Scientists agree that human activity is the primary driver of what we're seeing now world-wide. (This type of climate change is sometimes referred to as anthropogenic, which is just a way of saying "caused by human beings.") The unchecked burning of fossil fuels over the past 150 years has drastically increased the presence of atmospheric greenhouse gases, most notably carbon dioxide.

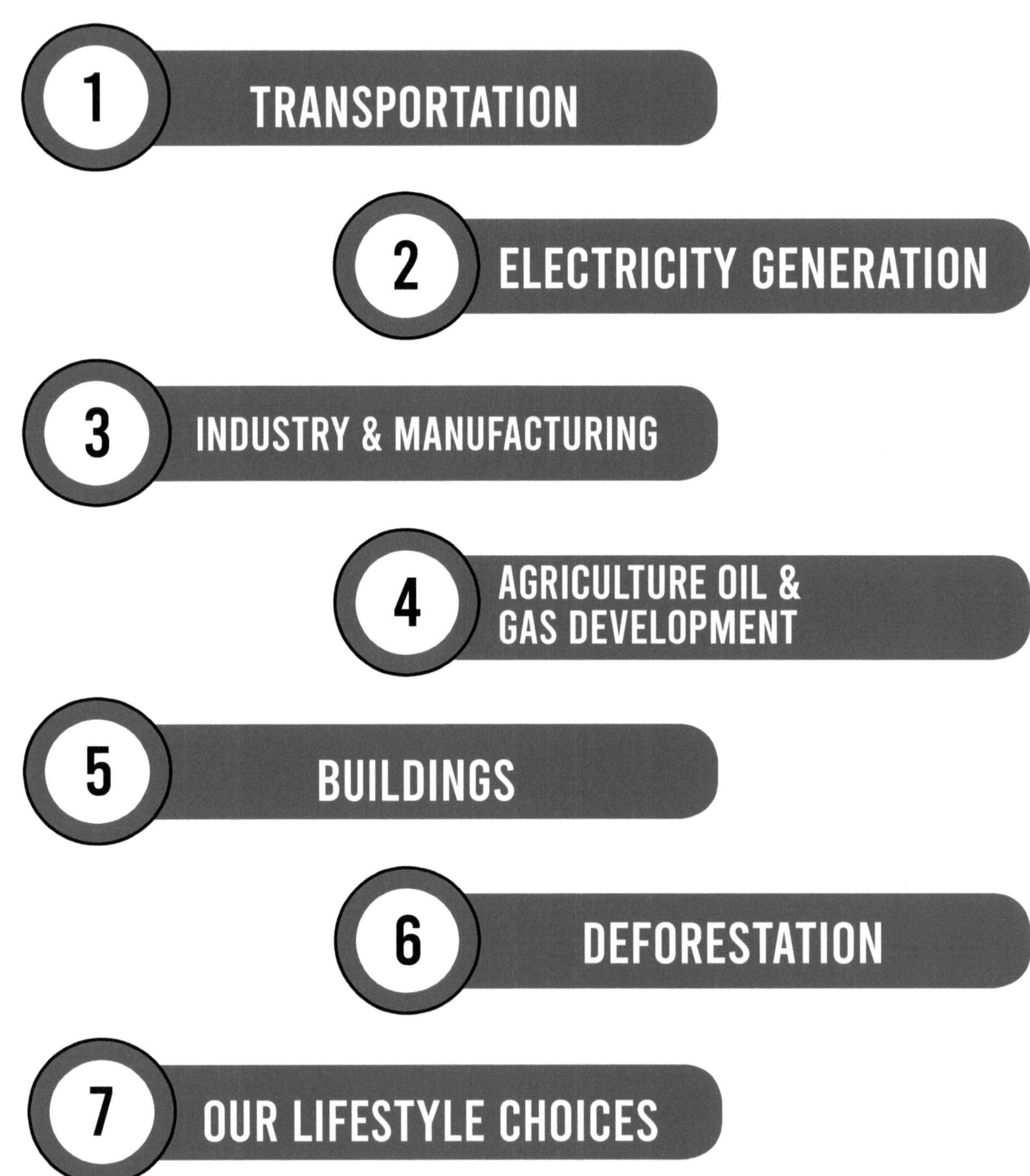

Source/2023-05-10-webinar-image.j:https://physicsworld.com/wp-content/uploads/2023/04pg
Reference link: https://www.nrdc.org/stories/what-are-causes-climate-change#natural

IMPACT OF CLIMATE CHANGE

Climate change is happening. Global average temperature has increased about 1.8°F from 1901 to 2016. Changes of one or two degrees in the average temperature of the planet can cause potentially dangerous shifts in climate and weather. These real, observable changes are what we call climate change impacts because they are the visible ways that climate change is affecting the Earth. For example, many places have experienced changes in rainfall, resulting in more floods, droughts, or intense rain, as well as more frequent and severe heat waves. The planet's oceans and glaciers have also experienced changes—oceans are warming and becoming more acidic, ice caps are melting, and sea level is rising. As these and other changes become more pronounced in the coming decades, they will likely present challenges to our society and our environment.

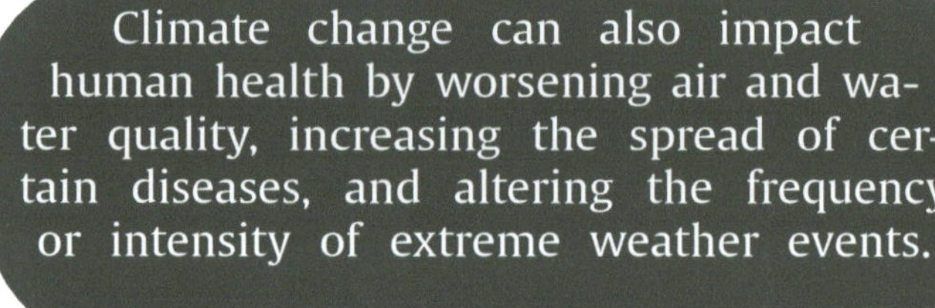

Warmer temperatures increase the frequency, intensity, and duration of heat waves, which can pose health risks, particularly for young children and the elderly.

Climate change can also impact human health by worsening air and water quality, increasing the spread of certain diseases, and altering the frequency or intensity of extreme weather events.

Changes in the patterns and amount of rainfall, as well as changes in the timing and amount of stream flow, can affect water supplies and water quality and the production of hydroelectricity and Rising sea level threatens coastal communities and ecosystems.

Changing ecosystems influence geographic ranges of many plant and animal species and the timing of their lifecycle events, such as migration and reproduction.

Increases in the frequency and intensity of extreme weather events, such as heat waves, droughts, and floods, can increase losses to property, cause costly disruptions to society, and reduce the affordability of insurance.

Source:https://science.nasa.gov/wp-content/uploads/2023/11/Effects_page_triptych.jpeg
Reference link:https://www.epa.gov/climatechange-science/impacts-climate-change

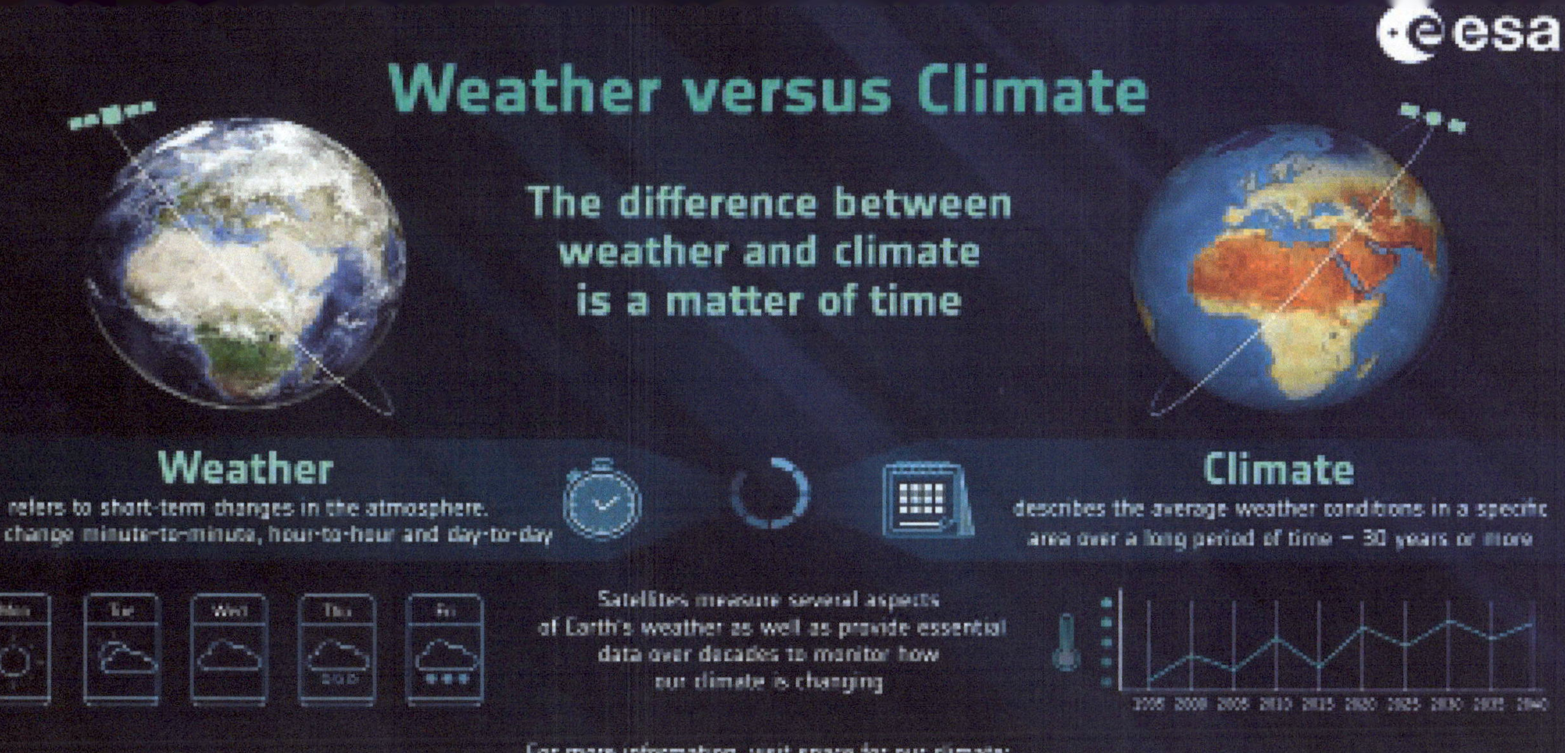

CLIMATE VS. WHETHER

WHAT'S THE DIFFERENCE ?

Although they are closely related, weather and climate are not the same. The difference between weather and climate is simply a matter of time. Weather refers to the short-term conditions of the atmosphere, while climate describes the average weather conditions over a long period of time.

WEATHER

Weather shows the way the atmosphere behaves and can change from minute-to-minute, hour-to-hour and day-to-day. There are many components to weather, which include temperature, rain, wind, hail, snow, humidity, flooding, thunderstorms, heatwaves and more. When you look outside your window on any given day, what you see is weather.

CLIMATE

Climate, on the other hand, is the weather in a specific area over a long period of time – usually 30 years of more. When scientists talk about climate, they look for trends or cycles of variability, such as changes in temperature, humidity, precipitation, ocean-surface temperature and other weather phenomena that occur over longer periods of time in a specific location.

Source:https://www.esa.int/ESA_Multimedia/Images/2021/04/Weather_versus_climate
Reference link:https://www.esa.int/Applications/Observing_the_Earth/Space_for_our_climate/Weather_vs_climate_What_s_the_difference

CLIMATE JUSTICE

Climate justice is a type of environmental justice that focuses on the unequal impacts of climate change on marginalized or otherwise vulnerable populations. Climate justice seeks to achieve an equitable distribution of both the burdens of climate change and the efforts to mitigate climate change. The economic burden of climate change mitigation is estimated by some at around 1% to 2% of GDP. Climate justice examines concepts such as equality, human rights, collective rights, justice and the historical responsibilities for climate change.

Climate justice recognizes that those who have benefited most from industrialization bear a disproportionate responsibility for the accumulation of carbon dioxide in the earth's atmosphere, and thus for climate change. Meanwhile, there is growing consensus that people in regions that are the least responsible for climate change as well as the world's poorest and most marginalized communities often tend to suffer the greatest consequences. Depending on the country and context, this may include people with low-incomes, indigenous communities or communities of color. They might also be further disadvantaged by responses to climate change which might exacerbate existing inequalities around race, gender, sexuality and disability. When those affected the most by climate change despite having contributed the least to causing it are also negatively affected by responses to climate change, this is known as the 'triple injustice' of climate change.

Fridays for Future demonstration in Berlin in September 2021 with the slogan "fight for climate justice

Source:https://www.orfonline.org/public/uploads/posts/image/1719514157_Climate.jpeg
Reference link: https://en.m.wikipedia.org/wiki/Climate_justice

CLIMATE SECURITY

Climate security is a political and policy framework that looks at the impacts of climate on security. Climate security often refers to the national and international security risks induced, directly or indirectly, by changes in climate patterns. It is a concept that summons the idea that climate-related change amplifies existing risks in society that endangers the security of humans, ecosystems, economy, infrastructure and societies. Climate-related security risks have far-reaching implications for the way the world manages peace and security. Climate actions to adapt and mitigate impacts can also have a negative effect on human security if mishandled. The term climate security was initially promoted by national security analysts in the US and later Europe, but has since been adopted by a wide variety of factors including the United Nations, low and middle income states, civil society organizations and academia. The term is used in fields such as politics, diplomacy, environment and security with increasing frequency. There are also critics of the term who argue that the term encourages a militarized response to the climate crisis, and ignores issues of misdistribution and inequity that underpin both the climate crisis and vulnerability to its impacts. Those who look at the national and international security risks argue that climate change has the potential to exacerbate existing tensions or create new ones – serving as a threat multiplier. For example, climate change is seen as a threat to military operations and national security, as the rise in sea level can affect military bases or extreme heat events can undermine the operability of armies. Climate change is also seen as a catalyst for violent conflict and a threat to international security, although the causality of climate and conflict is also debated. Due to the growing importance of climate security on the agendas of many governments, international organizations, and other bodies some now run programs which are designed to mitigate the effects of climate change on conflict. These practices are known as climate security practices. These practices stem from a variety of actors with different motivations in the sphere of development, diplomacy and defense; both NATO and the UN Security Council are involved in these practices.

Source:https://climatepromise.undp.org/sites/default/files/styles/large_2x/public/explainer/Climate%20Security%20-%201%20no%20logo.jpg?itok=ypxu0qlz

https://climatepromise.undp.org/news-and-stories/what-climate-security-and-why-it-important

CLIMATE OVERSHOOT

I n 2015 in Paris, the countries of the world agreed to limit global warming to well below 2 °C and to pursue efforts to limit it to 1.5 °C. In recent years, the lower goal has gained prominence, especially after the special report on global warming of 1.5°C from the Intergovernmental Panel on Climate Change (IPCC). The Climate Overshoot Commission understands "climate overshoot" to mean crossing the 1.5°C threshold.

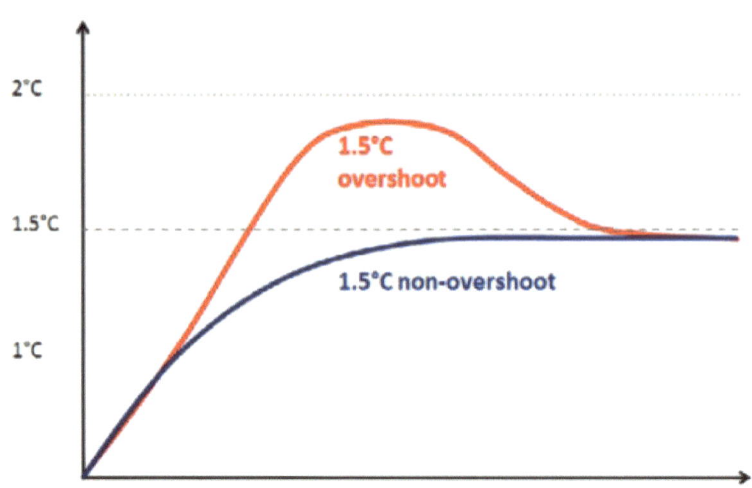

WHAT ARE THE RISKS OF CLIMATE OVER SHOOT?

Today's global warming of about 1.2°C is already creating significant impacts. The risks of overshooting the 1.5 °C goal will be felt across all the UN Sustainable Development Goals, with profound environmental and societal consequences. Beyond 1.5 °C there is a growing likelihood of "cascading and irreversible climate impacts". By 2 °C warming, many human and natural systems would be under extreme stress, with some ecosystems struggling to survive. At the same time, 1.5 °C and 2 °C are not absolute thresholds. Risks increase in line with the extent and duration of the overshoot. Every extra tenth of a degree and every extra decade of overshoot matter. This makes it all the more important to strive to reduce risks under all possible future conditions.

WHAT ARE THE APPROACHES TO REDUCE THE RISKS OF CLIMATE OVERSHOOT?

The Climate Overshoot Commission is examining four additional approaches. Each brings potential benefits, costs, limitations, and risks. They would affect people and regions differently and could be implemented poorly. They also raise issues for governance.

Source: https://i0.wp.com/granthaminstitute.com/wp-content/uploads/2023/11/overshoot.png?resize=640%2C352&ssl=1
Reference link: https://www.overshootcommission.org/overshoot

CLIMATE INDEXES

A climate index is here defined as a calculated value that can be used to describe the state and the changes in the climate system. The climate at a defined place is the average state of the atmosphere over a longer period of, for example, months or years. Changes on climate are much slower than on the weather that can change strongly day by day.

The first classical climate indices of the atmosphere have been defined already about approximately a hundred years ago, for example the North Atlantic Oscillation (NAO), the first found teleconnection pattern. The NAO index and the NAO pattern can be determined from monitoring station data or identified by means of EOF analysis. Climate indices allow a statistical study of variations of the dependent climatological aspects, such as analysis and comparison of time series, means, extremes and trends.

CLIMATE ELEMENTS AND CLIMATE INDICES

Each climate index is based on certain parameters and describes only certain aspects of the climate, so there are a variety of climate indices that have been defined and examined in numerous publications. For each climate index there is a defining equation that uses the so-called climate elements. These are measurable parameters that influence the properties of the climate system, primarily, for example, atmospheric parameters such as air pressure, air temperature, precipitation and solar radiation, but also non-atmospheric parameters such as sea surface temperature or ice cover. Typical climate indices that use air pressure are of a different type. Here matters not the absolute value of a station but the pressure gradient between locations. This means that at least two stations are needed, but in most cases field data are used to calculate atmospheric patterns and time series of climate indices. The same applies to the sea surface temperature. There are statistical measures used to describe and analyze climate patterns and variability. And, they are Climate base period, Climate indices for air temperature, Climate indices of precipitation, Climate indices for air pressure and Climate indices for sea surface temperature.

Source: https://www.un.org/sites/un2.un.org/files/2021/09/climatechange11.jpg
Reference link: https://www.cen.uni-hamburg.de/en/icdc/data/climate-indices.html

WHY WE NEED ACTION

Climate change is now affecting every country on every continent. It is disrupting national economies and affecting lives, costing people, communities and countries dearly today and even more tomorrow. People are experiencing the significant impacts of climate change, which include changing weather patterns, rising sea level, and more extreme weather events. The greenhouse gas emissions from human activities are driving climate change and continue to rise. They are now at their highest levels in history. Without action, the world's average surface temperature is projected to rise over the 21st century and is likely to surpass 3 degrees Celsius this century—with some areas of the world expected to warm even more.

WHY WE NEED ACTION

Affordable, scalable solutions are now available to enable countries to leapfrog to cleaner, more resilient economies. The pace of change is quickening as more people are turning to renewable energy and a range of other measures that will reduce emissions and increase adaptation efforts. But climate change is a global challenge that does not respect national borders. Emissions anywhere affect people everywhere. It is an issue that requires solutions that need to be coordinated at the international level and it requires international cooperation to help developing countries move toward a low-carbon economy. To address climate change, countries adopted the Paris Agreement at the COP21 in Paris on 12 December 2015. The Agreement entered into force less than a year later. In the agreement, all countries agreed to work to limit global temperature rise to well below 2 degrees Celsius, and given the grave risks, to strive for 1.5 degrees Celsius.

Source:https://www.watchdoguganda.com/wp-content/uploads/2024/01/Climate-Action.jpg
Reference link: https://www.un.org/sustainabledevelopment/climate-action/

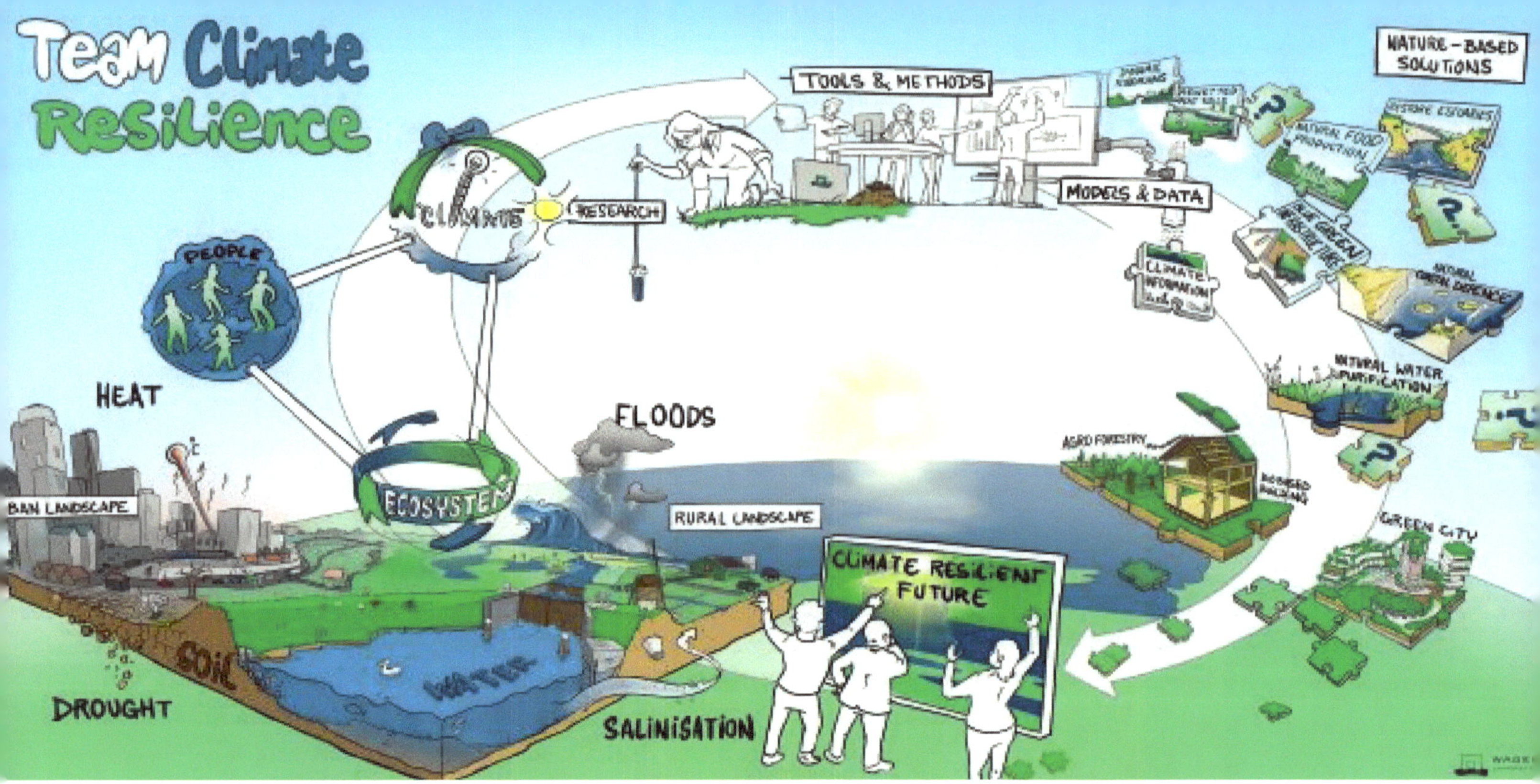

CLIMATE RESILIENCE

Climate resilience is about successfully coping with and managing the impacts of climate change while preventing those impacts from growing worse. A climate resilient society would be low-carbon and equipped to deal with the realities of a warmer world. The formal definition of the term is the "capacity of social, economic and ecosystems to cope with a hazardous event or trend or disturbance".

BUILDING CLIMATE RESILIENCE

To achieve climate resilience, we must combine mitigation and adaptation efforts, prioritizing climate justice. This will lead to a low-carbon society, equipped to handle warmer world realities, and safeguarding everyone's well-being.

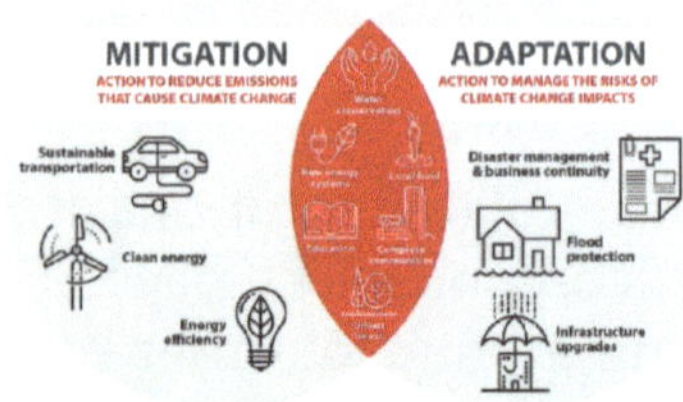

VISION STATEMENT OF CLIMATE RESILIENCE

By 2050 we all live in a 1.5-degree warmer world where all regions, countries, cities, businesses communities and individuals THRIVE in the face of multiple risks, uncertainty and threats posed by climate change. This vision of climate resilience is to be achieved through three interdependent outcomes

1 Resilient people and livelihoods

2 Resilient Businesses and Economies

3 Resilient Environmental Systems

Source:https://www.wur.nl/upload_mm/8/8/9/c65e9258-cf1e-4b19-b088-dcf65cff9d81_climate%20resilience%20praatplaat_951c2690_670x376.jpg
Reference link: https://www.ucsusa.org/reSources/what-climate-resilience | https://unfccc.int/sites/default/files/reSource/ExecSumm_Resilience_0.pdf
https://media.licdn.com/dms/image/D4D12AQHBhBqYRaxckg/article-cover_image-shrink_720_1280/0/1687596358329?e=2147483647&v=beta&t=LkLFeojVDb00tRmpHnOYWy0jucmh-qajL1dIL9BCgrrs

1.5 °

CLIMATE MITIGATION

Climate change mitigation refers to any action taken by governments, businesses or people to reduce or prevent greenhouse gases, or to enhance carbon sinks that remove them from the atmosphere. These gases trap heat from the sun in our planet's atmosphere, keeping it warm. Since the industrial era began, human activities have led to the release of dangerous levels of greenhouse gases, causing global warming and climate change. However, despite unequivocal research about the impact of our activities on the planet's climate and growing awareness of the severe danger climate change poses to our societies, greenhouse gas emissions keep rising. If we can slow down the rise in greenhouse gases, we can slow down the pace of climate change and avoid its worst consequences.

REDUCING GREENHOUSE GASES CAN BE ACHIEVED BY

- Shifting away from fossil fuels
- Improving energy efficiency
- Improving energy efficiency
- The sustainable management and conservation of forests
- Restoring and conserving critical ecosystems
- Creating a supportive environment

1.5°C GOAL AND ITS IMPORTANCE

The 1.5°C goal is extremely important, especially for vulnerable communities already experiencing severe climate change impacts. Limiting warming below 1.5°C will translate into less extreme weather events and sea level rise, less stress on food production and water access, less biodiversity and ecosystem loss, and a lower chance of irreversible climate consequences. To limit global warming to the critical threshold of 1.5°C, it is imperative for the world to undertake significant mitigation action. This requires a reduction in greenhouse gas emissions by 45 percent before 2030 and achieving net-zero emissions by mid-century.

Source: https://encrypted-tbn0.gstatic.com/images?q=tbn:ANd9GcQarDpQ4jZKdjEpENgkBg_6yQjey35jKESl2A&s
Reference link: https://climatepromise.undp.org/news-and-stories/what-climate-change-mitigation-and-why-it-urgent

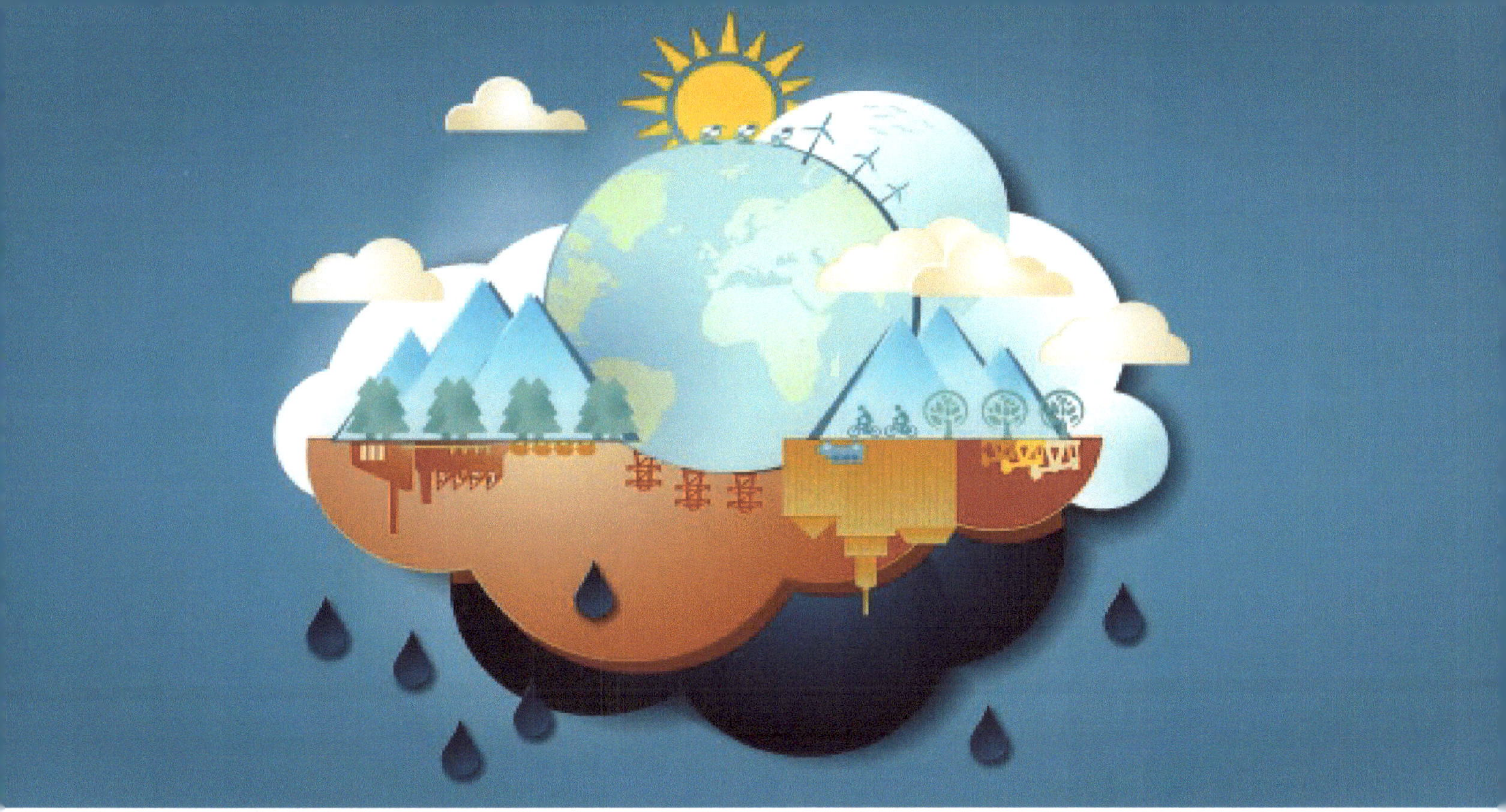

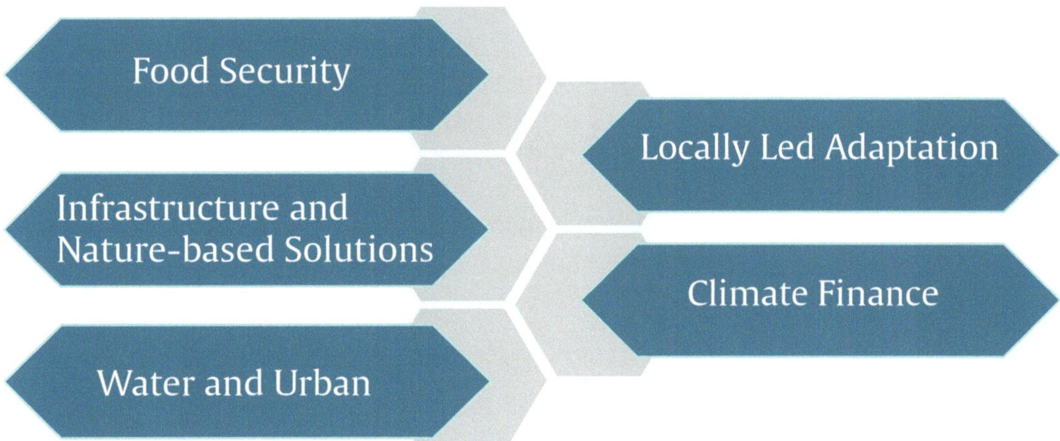

CLIMATE ADAPTATION

Climate adaptation means taking action to prepare for and adjust to the current and projected impacts of climate change. With climate change bringing more frequent and intense extreme weather events such as heat waves, droughts and floods, individuals and communities can reduce their vulnerability and increase their resilience by adapting now.

CLIMATE ADAPTATION CAN TAKE MANY DIFFERENT FORMS

- Food Security
- Locally Led Adaptation
- Infrastructure and Nature-based Solutions
- Climate Finance
- Water and Urban

ADAPTATION COMMITTEE

As part of the Cancun Adaptation Framework, Parties established the Adaptation Committee (AC) to promote the implementation of enhanced action on adaptation in a coherent manner under the

Convention. Since its establishment in 2010, the AC has become the United Nation's leading voice on adaptation. The sixteen-member body offers expert guidance on a range of adaptation topics, helping countries, civil society and businesses to build resilience and adapt to a changing climate,

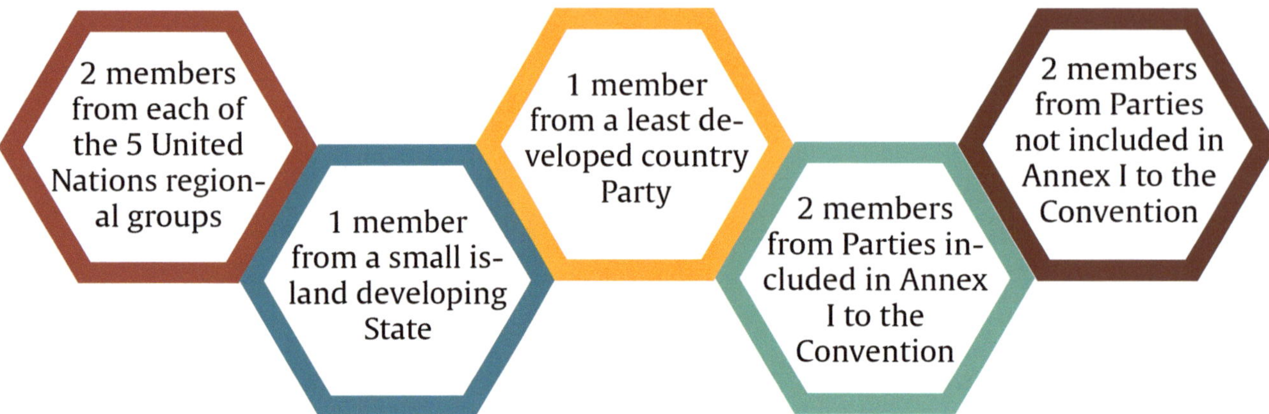

ADAPTATION FUND

The Adaptation Fund is an international fund that finances projects and programs aimed at help-ing developing countries to adapt to the harmful effects of climate change. It is set up under the Kyoto Protocol of the United Nations Framework Convention on Climate Change (UNFCCC).

The Adaptation Fund was officially launched in 2007, although it was established in 2001 at the 7th session of the Conference of the Parties (COP7) to the UNFCCC in Marrakech, Morocco to finance concrete adaptation projects and programmes that reduce the adverse effects of climate change facing communities, countries, and sectors. It is intended to finance climate adapta-tion projects and programmes in developing countries that are parties to the Kyoto Protocol.

CLIMATE FINANCE

Climate finance refers to local, national or transnational financing drawn from public, private and alternative Sources of financing that seeks to support mitigation and adaptation actions that will address climate change. The Convention, the Kyoto Protocol and the Paris Agreement call for financial assistance from Parties with more financial reSources to those that are less endowed and more vulnerable. This recognizes that the contribution of countries to climate change and their capacity to prevent it and cope with its consequences vary enormously. Climate finance is needed for mitigation, because large-scale investments are required to significantly reduce emissions. Climate finance is equally important for adaptation, as significant financial reSourc-es are needed to adapt to the adverse effects and reduce the impacts of a changing climate.

Source:https://i0.wp.com/oecd-development-matters.org/wp-content/uploads/2022/04/climate-adaptation-development-matters.jpg?fit=1200%2C700&ssl=1
Reference link: https://gca.org/ https://www4.unfccc.int/sites/NWPStaging/Pages/Adaptation-Committee.aspx https://en.wikipedia.org/wiki/The_Adaptation_Fund
https://unfccc.int/topics/introduction-to-climate-finance

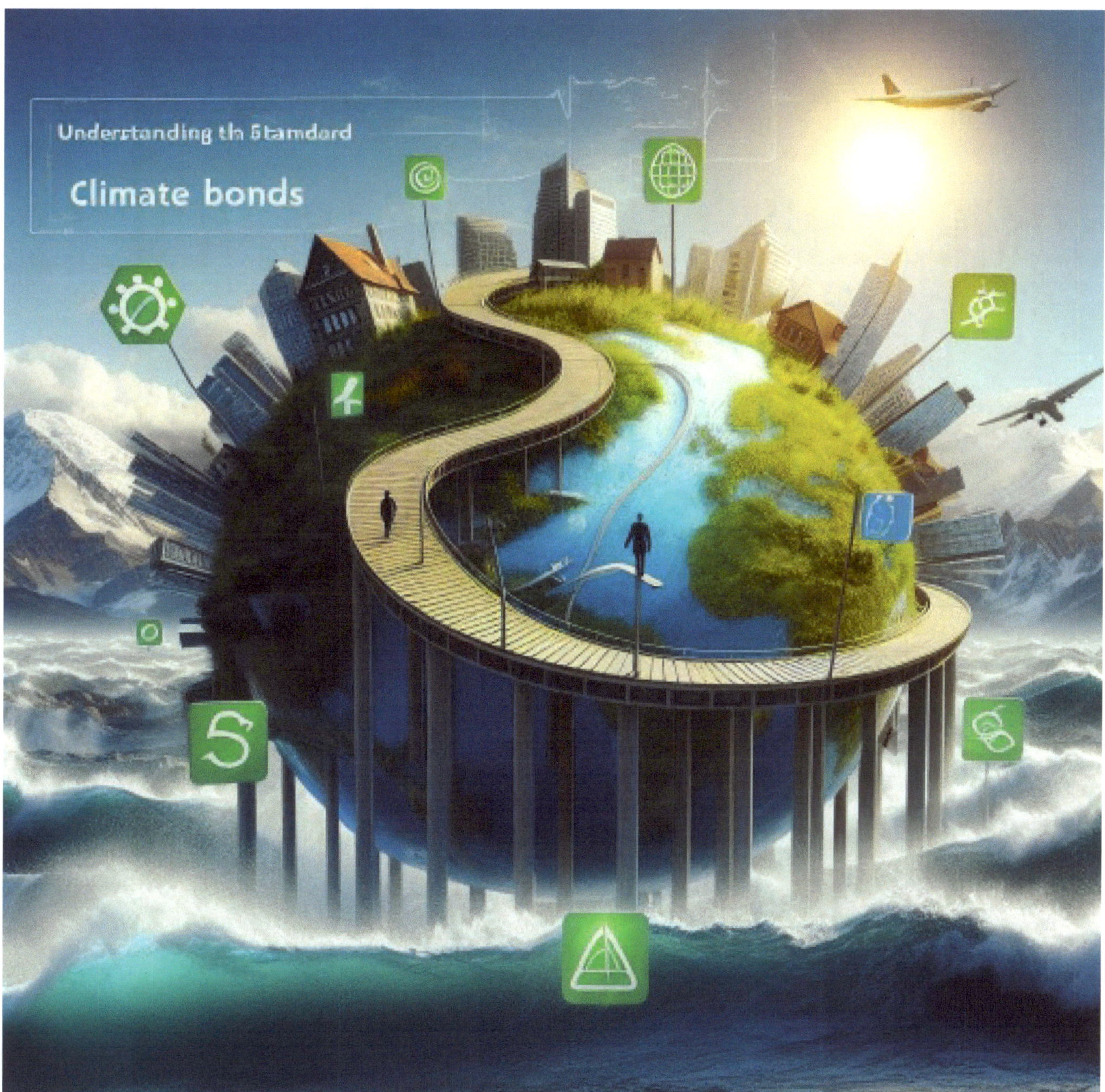

CLIMATE BOND

Climate bonds are fixed-income financial instruments (bonds) linked to climate change solutions. They are issued in order to raise finance for climate change solutions, for example mitigation or adaptation related projects. These might be greenhouse gas emission reduction projects ranging from clean energy to energy efficiency, or climate change adaptation projects ranging from building Nile delta flood defenses to helping the Great Barrier Reef adapt to warming waters. Like normal bonds, Climate Bonds can be issued by governments, multi-national banks or corporations. The issuing entity guarantees to repay the bond over a certain period of time, plus either a fixed or variable rate of return.

Most Climate Bonds are use-of-proceeds bonds, where the issuer promise to the investors that all the raised funds will only go to specified climate-related programs or assets, such as renewable energy plants or climate mitigation funding programs. We want investors to know that they are investing in climate change solutions.

SOME BOND TYPES ARE OBVIOUSLY USE-OF-PROCEEDS BONDS

PROJECT BONDS

where the money is in an separate company or special purpose vehicle (SPV) for a particular project.

ASSET-BACKED SECURITIES

where the money is for a portfolio of cash flows that are securitized in one bond, such as a portfolio of loans to renewable energy projects.

COVERED BONDS

where the investor has dual recourse to the issuer balance sheet (typically a bank) and also a pool of assets that are high quality (usually mortgages too).

Source:https://Cmiro.medium.com/v2/resize:fit:1024/1*mhxqJVb_vxK9xNP1RoWeig.jpeg
Reference link:
https://www.climatebonds.net/reSources/understanding

CLIMATE PACT

The Climate Pact provides a opportunity for individuals to learn about climate change, develop and execute solutions, and work together to maximize their effect. It including the European Climate Pact and the Glasgow Climate Pact

EUROPEAN CLIMATE PACT

The European Climate Pact is a movement of people united around a common cause, each taking steps in their own worlds to build a more sustainable Europe. Launched by the European Commission, the Pact is part of the European Green Deal and is helping the EU to meet its goal to become climate-neutral by 2050.

The European Climate Pact encourages everyone to act. It is a movement of people united around a common cause, each taking steps in their own worlds to build a more sustainable Europe for us all. Launched by the European Commission, the Pact is part of the European Green Deal and is helping the EU to meet its goal to be the first climate-neutral continent in the world by 2050.

GLASGOW CLIMATE PACT

Nations adopted the Glasgow Climate Pact, aiming to turn the 2020s into a decade of climate action and support. The package of decisions consists of a range of agreed items, including strengthened efforts to build resilience to climate change, to curb greenhouse gas emissions and to provide the necessary finance for both. Nations reaffirmed their duty to fulfill the pledge of providing 100 billion dollars annually from developed to developing countries. And they collectively agreed to work to reduce the gap between existing emission reduction plans and what is required to reduce emissions, so that the rise in the global average temperature can be limited to 1.5 degrees. For the first time, nations are called upon to phase down unabated coal power and inefficient subsidies for fossil fuels.

Source:https://img.etimg.com/thumb/width-1200,height-900,imgsize-58173,resizemode-75,m-
sid-50159798/news/international/world-news/world-leaders-welcome-historic-climate-pact.jpg
Reference link:
https://climate-pact.europa.eu/index_en
https://unfccc.int/process-and-meetings/the-paris-agreement/the-glasgow-climate-pact-key-
outcomes-from-cop26
https://www.eesc.europa.eu/sites/default/files/styles/large/public/images/shutter-
stock_1339316864.jpg?itok=Jhi7-oad

CLIMATE AMBITION

Climate ambition is a commitment and plans to shape the future. It's a positive vision for a net zero world, for people, business, economy and environment. Ambition guides strategy, empowers employees, and informs stakeholders.

AIM FOR ZERO

Set science-based net zero targets with near-term and long-term goals, in line with a 1.5°C pathway. This means setting targets to halve emissions by 2030 and reach net zero no later than 2050 – ideally by 2040.

FACTOR IN NATURE

Prioritize nature-based solutions to build a net zero, nature-positive future faster.

CLIMATE AMBITION SUMMIT

20 SEPTEMBER 2023, UNITED NATIONS HEADQUARTERS, NEW YORK

Against the backdrop of the worsening climate crisis, the UN Secretary-General's Climate Ambition Summit aimed to showcase "first mover and doer" leaders from government, business, finance, local authorities, and civil society who came with credible actions, policies and plans – and not just pledges – to accelerate the decarbonization of the global economy and deliver climate justice in line with his Acceleration Agenda.

"If these first-doers and first-movers can do it, everybody can do it," the Secretary-General said in his closing remarks, calling it a "Summit of Hope."

The design and outcomes of the Summit will be delivered on three distinct but interrelated acceleration tracks are ambition, credibility and implementation.

Source:https://sgkplanet.com/wp-content/uploads/2021/09/FAQs-about-Climate-Ambition-Website-SGK-PLANET-Sandor-Alejandro-Gerendas-Kiss-T.jpg
Reference link: https://www.wemeanbusinesscoalition.org/ambition/
https://www.un.org/en/climatechange/climate-ambition-summit

CLIMATE NEGOTIATION

The climate negotiation process occurring through the United Nations Framework Convention on Climate Change (UNFCCC) and its related agreements is the primary forum for international cooperation on stabilizing atmospheric greenhouse gas concentrations at a level that would prevent catastrophic anthropogenic interference with the climate system. The Kyoto Protocol to the UNFCCC, adopted in 1997 and entered into force in 2005, imposes emissions reduction and limitation obligations on industrialized country Parties. The Paris Agreement to the UNFCCC, which was adopted in 2015 and rapidly entered into force in 2016, commits all states to take climate action on the basis of equity and to keep global temperature increase below 1.5°C.

The UNFCCC negotiation process is focused on long-term cooperative action to address climate change. This shared vision requires ambitious mitigation and adaptation action coupled with enhanced financial support to make it possible. Human rights are affected by both climate change itself and actions to combat it, making them vital to address within the UNFCCC. The Paris Agreement explicitly recognized that the existing human rights obligations of Parties to the Paris Agreement should be fully respected, promoted, and considered in all climate change-related activities.

The "UNFCCC Story" gallery has a section related to the negotiation process, with artifacts related to those processes. Artifacts include original credentials issued by Parties, a replica of the green leaf gavel from COP21, and draft negotiating texts and original submissions from Parties.

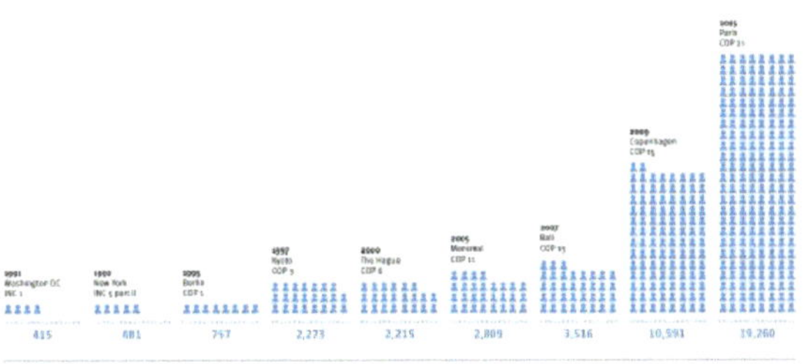

Source:https://img.freepik.com/premium-photo/hand-power-cooperation-businessto-succeed-developing-sustainable-partnershipsconcept-ai-generated_1261112-910.jpg?size=626&ext=jpg&ga=GA1.1.1826414947.1723334400&semt=ais_hybrid Reference link: https://unfccc.int/about-us/unfccc-archives/the-unfccc-archival-exhibition/the-global-negotiation-process
https://www.ciel.org/issue/climate-negotiations/

CLIMATE EMPOWERMENT

Climate empowerment is a term used by the United Nations Framework Convention on Climate Change (UNFCCC) to describe the work done under Article 6 of the Convention and Article 12 of the Paris Agreement. The goal of climate empowerment is to empower people to take action on climate change.

ACTION FOR CLIMATE EMPOWERMENT

Action for Climate Empowerment (ACE) is a term adopted by the UN Framework Convention on Climate Change to denote work under Article 6 of the Convention and Article 12 of the Paris Agreement. The over-arching goal of ACE is to empower all members of society to engage in climate action, through the six ACE elements - climate change education and public awareness, training, public participation, public access to information, and international cooperation on these issues.

ACE MATTERS

Implementation of all six ACE elements is crucial to the global response to climate change. Everyone, including and perhaps especially the young, must understand and participate in the transition to a low-emission, climate-resilient world.

THE SIX ELEMENTS OF ACE

ACE addresses all six priority areas of Article 6 Education, Training, Public awareness, Public access to information, Public participation, and International cooperation.

ACE AT COP27

Parties adopted a four-year ACE action plan under the Glasgow work programme (Decisions 23/CP.27 and 22/CMA.4), which sets out short-term, clear and time-bound activities in the four priority areas of the work programme and across the six elements of ACE in a balanced manner as a concrete step towards empowering all members of society, including children and youth, to engage in climate action.

Source: https://upprojects.imgix.net/images/Climate-empowerment-web-image.png?fm=auto&auto=format&q=80&w=640&
Reference link https://unfccc.int/topics/education-and-youth/big-picture/ACE
https://en.wikipedia.org/wiki/Action_for_Climate_Empowerment

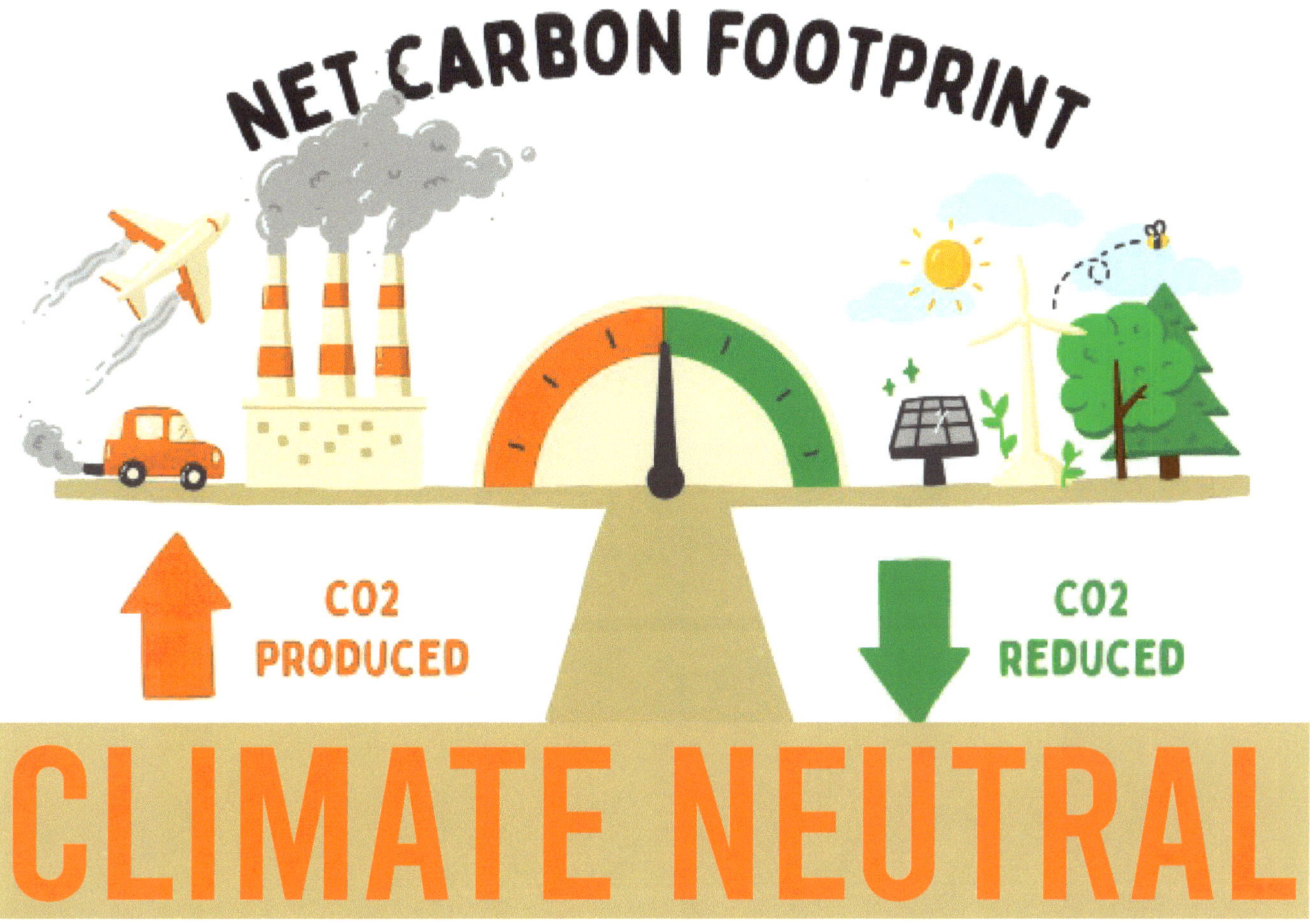

Climate neutrality refers to the idea of achieving net zero greenhouse gas emissions by balancing those emissions so they are equal to, or less than, the emissions removed, as well as accounting for regional or local biogeophysical effects of human activities, such as changes in surface albedo or local climate. In basic terms, it means we reduce our emissions through climate action to ensure no net effect on the climate system. As part of this, UN Climate Change launched Climate Neutral Now back in 2015 in order to encourage stakeholders around the world to work towards net zero emissions and a climate neutral world. While the aim is to have a 'climate neutral' world by 2050, Climate Neutral Now focuses on the need to take action now in order to reach that target.

CARBON NEUTRAL DOES NOT MEAN CARBON FREE

The terms carbon free and carbon neutral are often confused, however, they refer to distinct aspects of climate action. Carbon-free products, services or companies are those that do not generate any carbon emissions during the manufacturing, provision or operational process. This must apply to the entire supply chain, including all the raw materials, logistics and packaging. In actual fact, there are no examples (yet) of carbon-free products. Conversely, any company and any product can be carbon neutral there are current standards to calculate their emissions, and companies can support certified carbon offset projects in order to offset the emissions calculated.

Source: https://teepublic.zendesk.com/hc/article_attachments/5771949859479
Reference link: https://unfccc.int/news/a-beginner-s-guide-to-climate-neutrality
https://www.climatepartner.com/en/knowledge/glossary/carbon-neutral

CLIMATE POSITIVE ADVOCACY

Climate positive advocacy is an approach that goes beyond merely addressing carbon emissions to actively improving the environment. This concept involves creating a net-positive impact on the climate by not just reducing greenhouse gases but also by enhancing ecosystems and fostering environmental health.

HERE, ARE SOME KEY ASPECTS

CARBON REMOVAL	This includes initiatives that capture and store more carbon dioxide from the atmosphere than is emitted. Examples are reforestation, afforestation, and carbon capture and storage technologies.

ECOSYSTEM RESTORATION	Supporting projects that restore degraded ecosystems, such as wetlands, forests, and mangroves, which can sequester carbon and enhance biodiversity.
SUSTAINABLE PRACTICES	Advocating for and adopting practices that have a positive impact on the environment, such as sustainable agriculture, green building practices, and renewable energy use.
POLICY ADVOCACY	Promoting policies that drive systemic changes towards a climate-positive future, such as stricter emissions regulations, incentives for green technology, and support for conservation efforts.
EDUCATION AND AWARENESS	Raising awareness about climate-positive actions and their benefits, and encouraging individuals and organizations to adopt these practices.

Overall, climate positive advocacy is about taking proactive steps to not just mitigate climate change but actively contribute to a healthier, more resilient planet.climate positive advocacy represents a forward-thinking approach to tackling climate change, aiming not only to mitigate negative impacts but also to actively improve the environment and climate. By embracing and promoting carbon removal, ecosystem restoration, sustainable practices, supportive policies, and widespread education, we can work towards a future where our collective efforts lead to a net-positive impact on the planet.

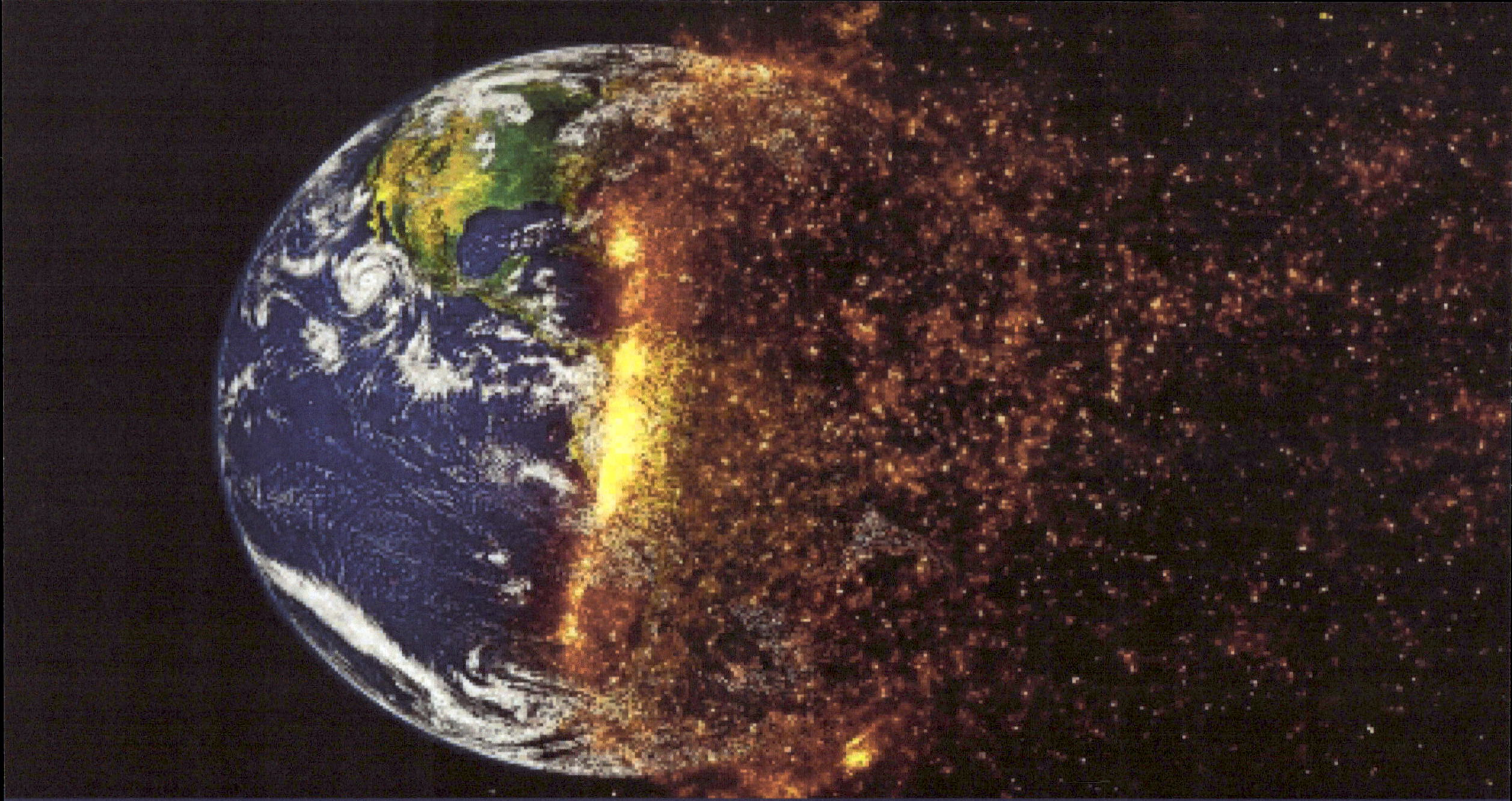

CLIMATE CRISIS

Climate crisis is a term that is used to describe global warming and climate change, and their effects. This term and the term climate emergency have been used to describe the threat of global warming to humanity and Earth, and to urge aggressive climate change mitigation and transformational adaptation. Nature and climate risks are getting the attention they deserve that's a positive first step in addressing some of the greatest challenges that we, as a global community, face. Scientists announced that temperatures in 2023 reached 1.48°C above preindustrial averages, with the 1.5°C threshold that takes the Earth into an unsafe operating space likely to be breached in the next 12 months.

The World Economic Forum's Global Risks Report 2024 named three key climate issues as critical challenges facing humanity - Extreme weather events, critical change to Earth systems which is a new entrant this year and biodiversity loss and ecosystem collapse.These climate-related risks sit alongside challenges like disinformation, geo-political competition and inflation as the definitive risks of 2024 and beyond.

CAUSES FOR RISING EMISSIONS

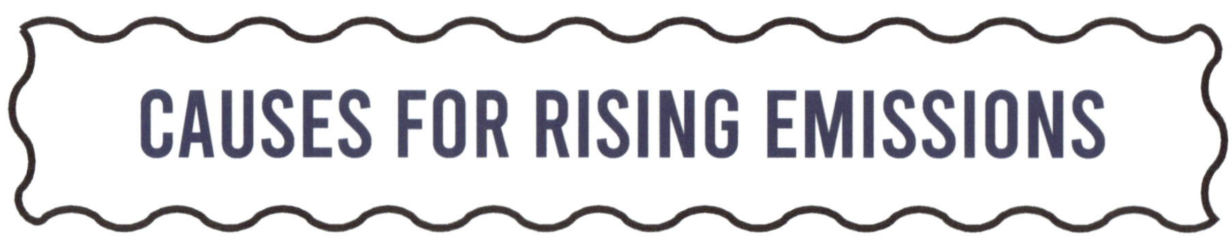

1 Burning coal, oil and gas produces carbon dioxide and nitrous oxide.

2 Cutting down forests (deforestation). Trees help to regulate the climate by absorbing CO_2 from the atmosphere. When they are cut down, that beneficial effect is lost and the carbon stored in the trees is released into the atmosphere, adding to the greenhouse effect.

3 Increasing livestock farming. Cows and sheep produce large amounts of methane when they digest their food.

4 Fertilizers containing nitrogen produce nitrous oxide emissions.

5 Fluorinated gases are emitted from equipment and products that use these gases. Such emissions have a very strong warming effect, up to 23 000 times greater than CO_2.

CLIMATE VARIATION

C limate variation refers to the complex and multifaceted problem of changes in climate patterns, including the interplay of societal, political, economic, scientific, and ethical factors. It is a global issue that can have long-lasting effects at both local and global scales, with the Earth taking a significant amount of time to adjust to warming due to the lingering presence of greenhouse gases like CO_2 in the atmosphere.

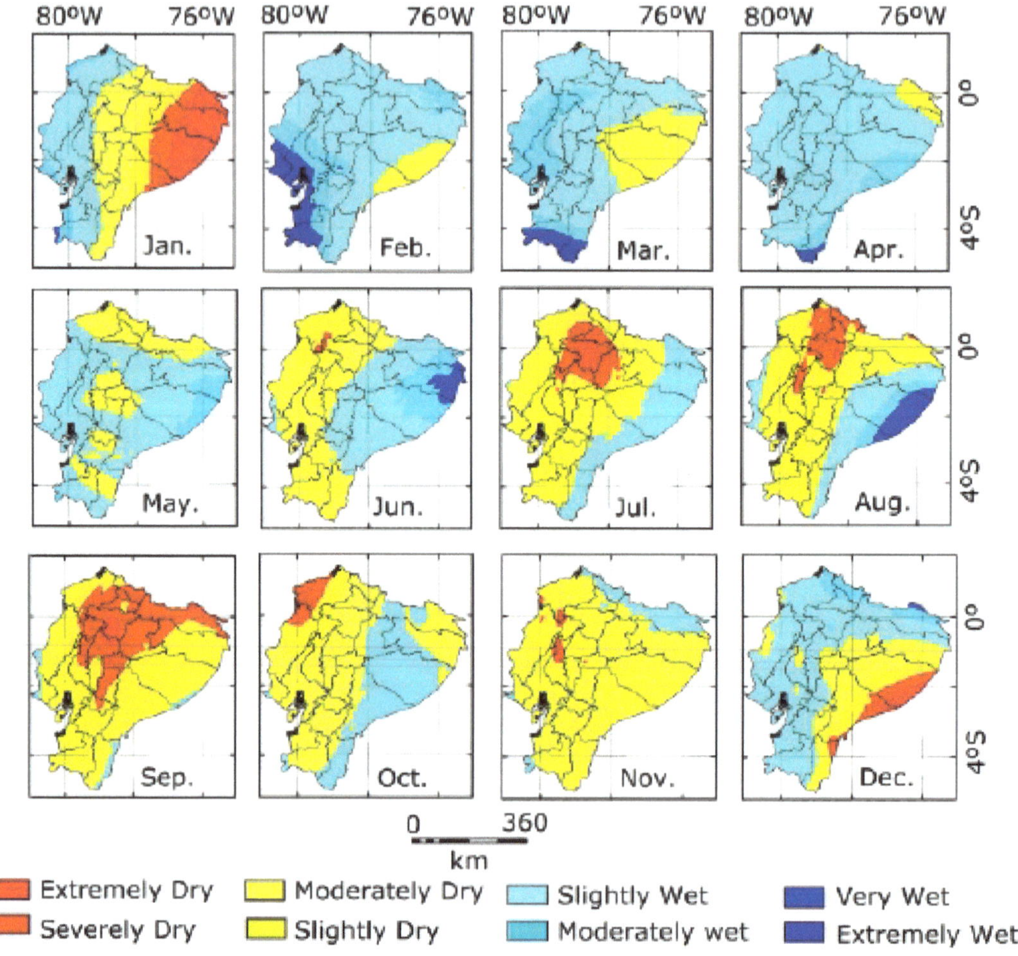

 Weather can be highly variable on a daily, weekly, or even yearly basis one day it might be sunny and the next it's snowing. Climate, on the other hand, doesn't change day-to-day because it is based on longer time scales and averages. However, climate is variable as well.

 Climate variability is the way aspects of climate (such as temperature and precipitation) differ from an average. Climate variability occurs due to natural and sometimes periodic changes in the circulation of the air and ocean, volcanic eruptions, and other factors.Climate variability is often natural, however climate change is causing an increase in the probability of many extreme weather events, and those events contribute to climate variability.

Climatologists are concerned with more than temperature changes as climate changes. Extreme precipitation events are also important. Precipitation patterns that deviate significantly from the average can result in droughts or floods. The flooding in Texas in August 2017 caused by Hurricane Harvey is a recent and devastating example of an extreme precipitation event.

Source:https://media.springernature.com/lw685/springer-static/image/art%3A10.1007%2Fs11269-020-02549-w/MediaObjects/11269_2020_2549_Fig7_HTML.png
Reference link: https://scied.ucar.edu/learning-zone/how-climate-works/climate-variability#:~:text=However%2C%20climate%20is%20variable%20as,volcanic%20eruptions%2C%20and%20other%20factors

A climate emergency is a situation in which urgent action is required to reduce or halt climate change and its effects. Here's a detailed explanation: A climate emergency is a state of crisis caused by human-induced climate change, characterized by severe and far-reaching consequences for the environment, human health, and the economy.

Causes

1. Greenhouse gas emissions: The burning of fossil fuels (coal, oil, and gas) and land use changes (deforestation, agriculture) release large amounts of carbon dioxide and other greenhouse gases, leading to global warming.
2. Deforestation and land degradation: The clearance of forests for agriculture, urbanization, and other purposes releases carbon stored in trees and reduces the ability of forests to act as carbon sinks.
3. Population growth and consumption patterns: As the global population grows, so does energy demand, leading to increased greenhouse gas emissions.

Effects

1. Rising global temperatures: The average global temperature has risen by about 1°C since the late 1800s, leading to more frequent and severe heatwaves, droughts, and storms.
2. Extreme weather events: Climate change is linked to an increase in extreme weather events, such as hurricanes, wildfires, and floods.
3. Sea-level rise: Thawing of polar ice caps and glaciers, and the expansion of seawater as it warms, are causing sea levels to rise, leading to coastal erosion and flooding.
4. Water scarcity: Changes in precipitation patterns and increased evaporation due to warmer temperatures are leading to droughts and water scarcity in many regions.
5. Loss of biodiversity: Climate change is altering ecosystems, leading to the loss of biodiversity, and extinction of many plant and animal species.

Consequences

1. Human migration and conflict: Climate change is expected to lead to the displacement of millions of people, particularly in vulnerable communities, and may exacerbate social and economic tensions.
2. Food insecurity: Climate change is impacting agricultural productivity, leading to crop failures, reduced yields, and changed growing seasons.

Source: https://coveringclimatenow.org/wp-content/uploads/2021/11/climate-emergency-protest-sign.jpeg Reference link: https://www.unep.org/climate-emergency
https://en.wikipedia.org/wiki/Climate_emergency_declaration

3. Economic disruption: Climate change is expected to have significant economic impacts, from damage to infrastructure and increased healthcare costs to impacts on tourism and recreation.

Urgent Action Required

The climate emergency demands immediate attention and collective action to:
1. Reduce greenhouse gas emissions: Transition to renewable energy sources, increase energy efficiency, and electrify transportation.
2. Protect and restore natural carbon sinks: Preserve and expand forests, wetlands, and other ecosystems that absorb carbon dioxide.
3. Develop climate-resilient infrastructure: Invest in infrastructure that can withstand the impacts of climate change, such as sea-level rise and extreme weather events.
4. Support climate change research and development: Continuously fund and conduct research to improve our understanding of climate change and develop effective solutions.

The window for action is rapidly closing. We must work together to mitigate the effects of climate change and create a sustainable future for all. The term climate emergency has been promoted by climate activists and pro-climate action politicians to add a sense of urgency for responding to a long-term problem. A United Nations Development Programme survey of public opinion in 50 countries found that sixty-four percent of 1.2 million respondents believe climate change is a global emergency.

Climate emergency as a term was used in protests against climate change before 2010 (e.g. the "Climate-Emergency-Rally" in Melbourne in June 2009. In 2017 the city council of Darebin adopted multiple measures named "Darebin Climate Emergency Plan". On 4 December 2018, the Club of Rome presented their "Climate Emergency Plan", which included 10 high-priority measures to limit global warming. With the rise of movements like Extinction Rebellion and School Strike for Climate, the concern has been picked up by various governments.

Multiple European cities and communities who declared a climate emergency are simultaneously members of the Klima-Bündnis (German for climate alliance), which obligates them to lower their CO_2 emissions by 10% every five years.

Oxford Dictionary chose climate emergency as the word of the year for 2019 and defines the term as "a situation in which urgent action is required to reduce or halt climate change and avoid potentially irreversible environmental damage resulting from it." Usage of the term soared more than 10,000% between September 2018 and September 2019.

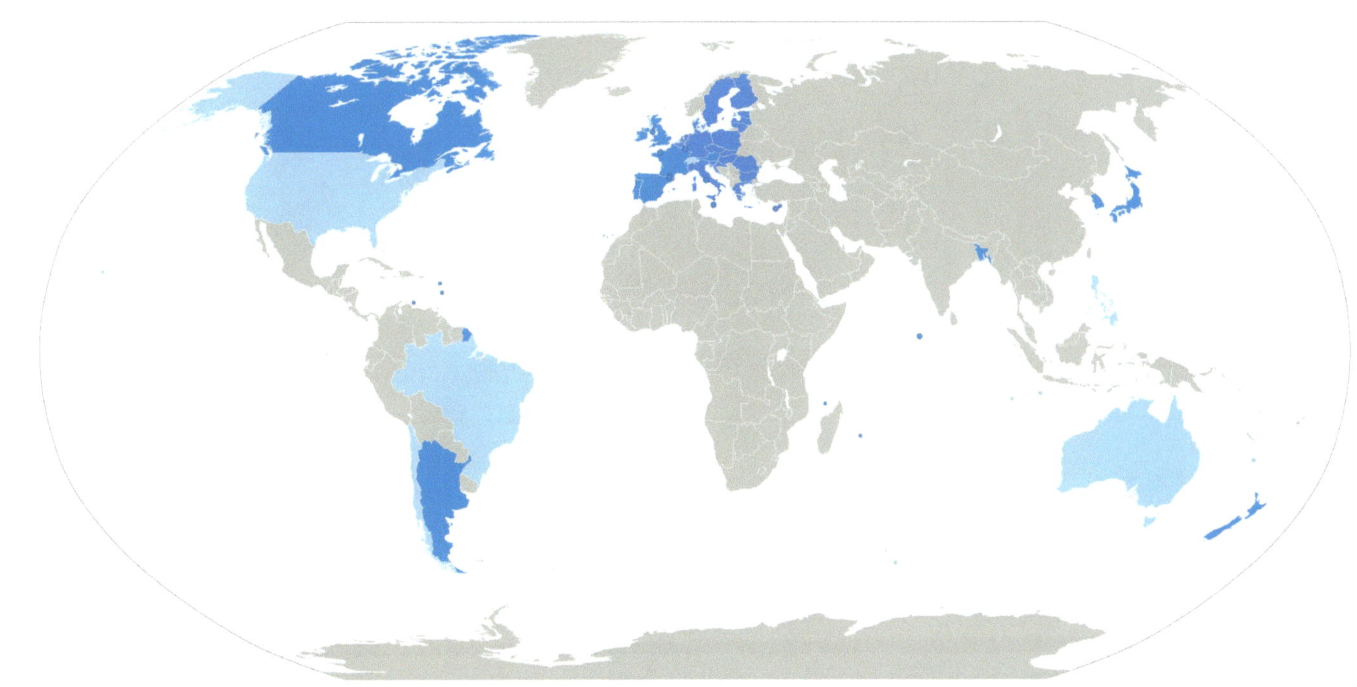

LIMATE
MERGENCY
RISEANDRESIST.ORG

What is climate fintech?

Climate Fintech Digital financial technology that catalyses decarbonization. Climate Fintech is simply the intersection of climate, finance, and digital technology. These digital innovations, applications, and platforms serve as crucial financial intermediaries and mediums between all stakeholders pursuing decarbonization.

The inaugural Climate Fintech Report explores crucial intersections of digital financial technology and climate as a fresh perspective by which to pursue decarbonization – through nurturing an emerging digital ecosystem of climate capital catalysts. Fintech can help mobilize more capital in the pursuit of reducing greenhouse gas emissions (GHG). The inaugural Climate Fintech Report explores crucial intersections of digital financial technology and climate as a fresh perspective by which to pursue decarbonization – through nurturing an emerging digital ecosystem of climate capital catalysts.

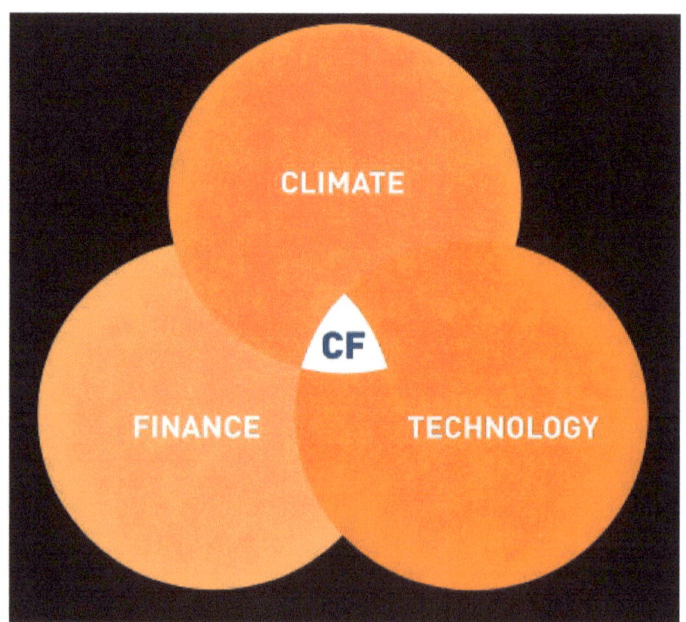

Source: https://www.pragmaticcoders.com/wp-content/uploads/2024/07/What-is-climate-fintech_-optimized.jpg
Reference link: https://www.climatefintech.cn/#what-is-climate-fintech

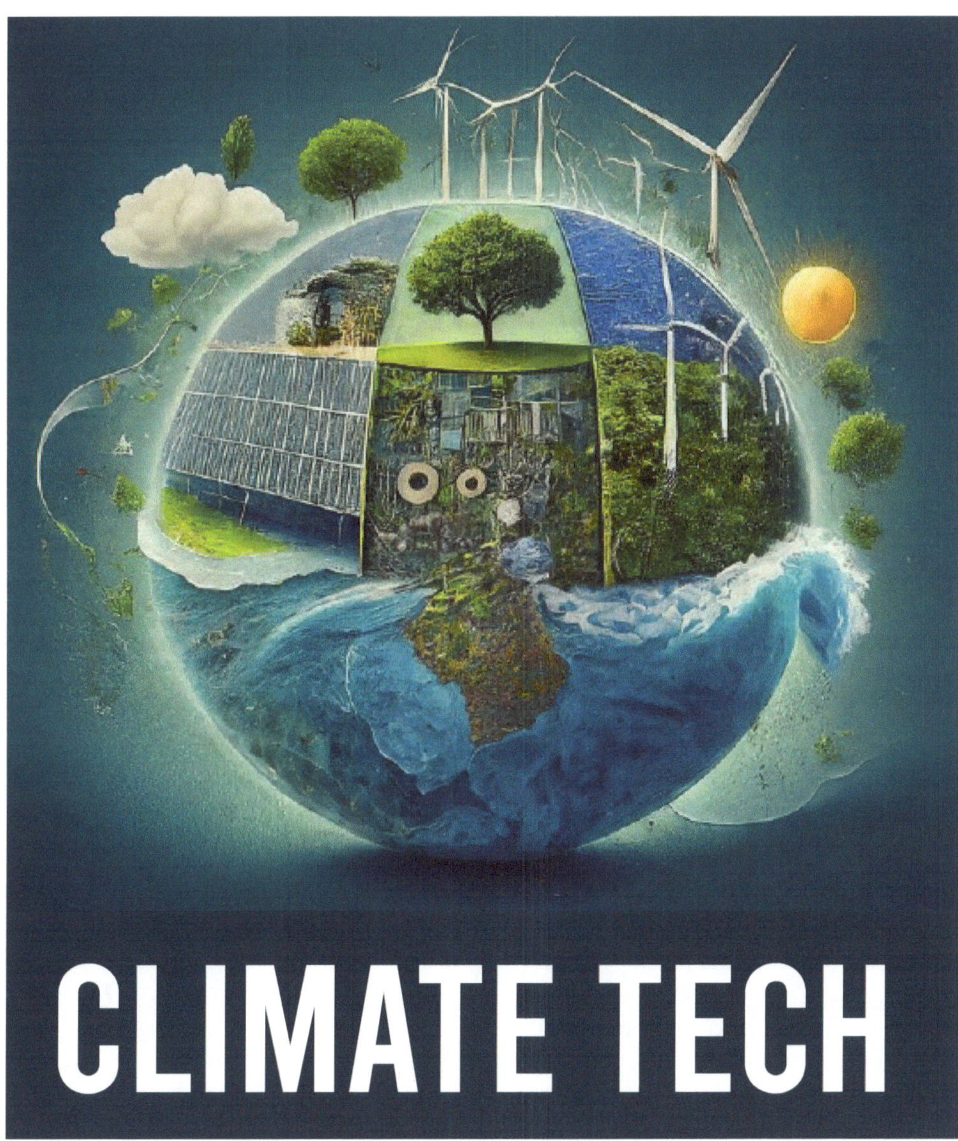

Climate tech refers to technologies and services that enable decarbonisation of the global economy. Climate tech companies develop products and services that leverage these technologies to mitigate and adapt to climate change, by removing existing carbon from the atmosphere, reducing future emissions or by increasing our resilience against the impacts of a changing climate. Since addressing climate change requires transformation across all sectors, climate tech companies span a wide variety of end markets and business models. he technologies addressing the five main emissions sectors Energy, Transportation, Industry, Building, Agricultureare defined by the IPCC, a sixth sector has also emerged – "carbon tech."

This category encompasses two main types of innovation: carbon removal technology and carbon accounting-related software. Carbon measurement and management platforms are often classified as an 'enabling' climate tech sector as they support the success of solutions in the other five emissions categories. Carbon removal entered the climate innovation spotlight in recent years, especially after the IPCC stated in its 2022 report that "the deployment of CDR to counterbalance hard-to-abate residual emissions is unavoidable if net zero CO_2 or GHG emissions are to be achieved"Common technology types include Direct Air Capture (DAC), Biochar, Soil Carbon Sequestration, Enhanced Weathering, and Bioenergy with Carbon Capture and Storage (BECCS), some of which are capital intensive and mostly still in the R&D phase without proven commercial success.

The term climate tech grew in popularity beginning around 2019. While climate tech still includes renewable energy and electric vehicle start-ups, since the core technology has already proven itself, new entrants into these markets now must provide a differentiated value add related to either scaling existing solutions or increasing the capacity and efficiency of existing technologies. Climate tech also includes new climate-specific sectors that did not exist during the clean tech investing era, most notably carbon tech.

Source: https://media.licdn.com/dms/image/D4D12AQEl1ohcKv7AOA/article-cover_image-shrink_720_1280/0/1707135147134?e=2147483647&v=beta&t=pOU8kKva3tFk8Rwdrap8a8O1CUh-2hiIssc_SpOxTul8 Reference link: https://www.nasdaq.com/solutions/listings/reSources/blogs/what-is-climate-tec

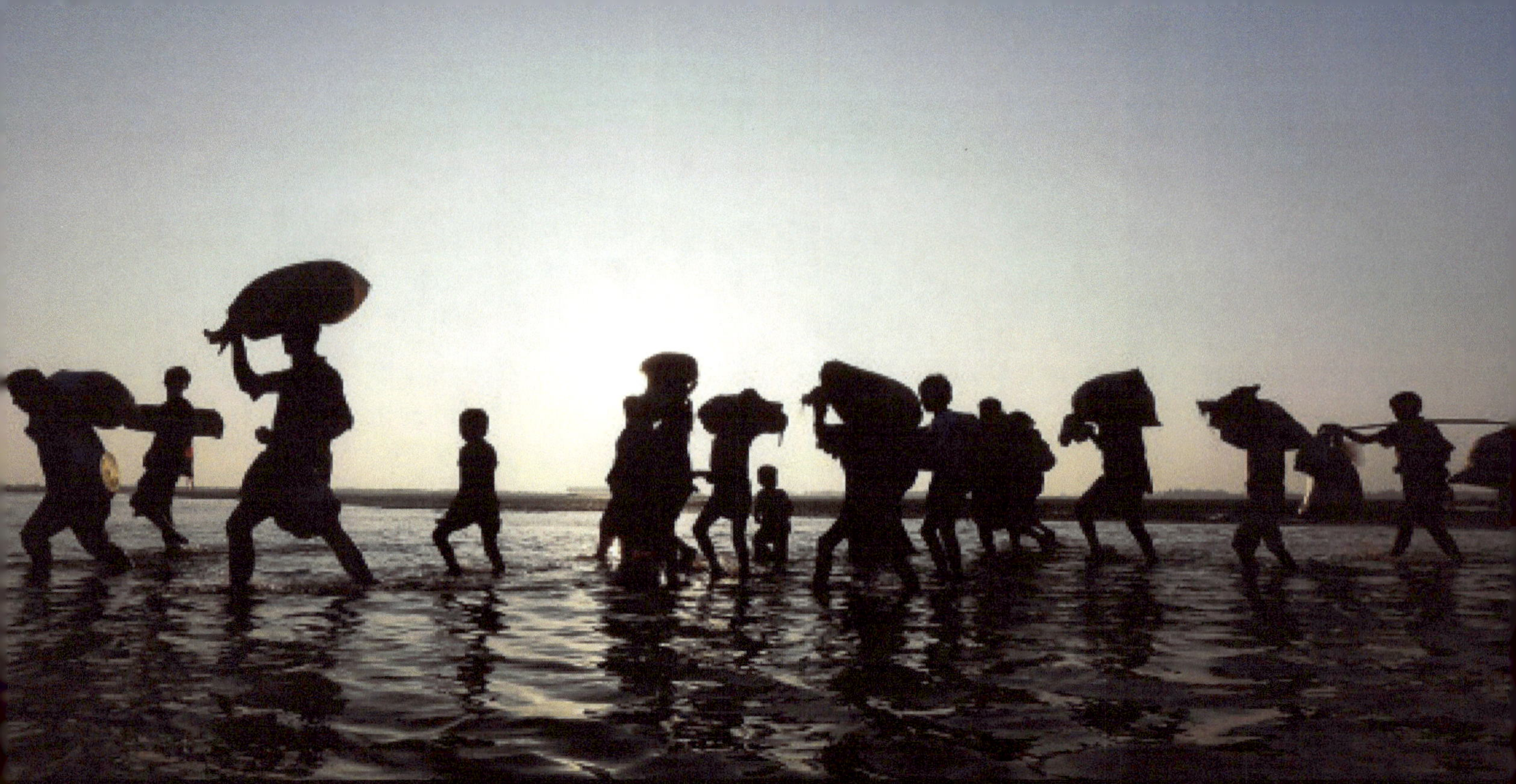

CLIMATE TECH

According to recent statistics published by the Internal Displacement Monitoring Centre, over 376 million people around the world have been forcibly displaced by floods, windstorms, earthquakes or droughts since 2008, with a record 32.6 million in 2022 alone. Since 2020, there has been an annual increase in the total number of displaced people due to disaster compared with the previous decade of 41 % on average.

The upward trend is alarmingly clear. With climate change as the driving catalyst, the number of 'climate refugees' will continue to rise. The Institute for Economics and Peace predicts that in the worst-case scenario, 1.2 billion people could be displaced by 2050 due to natural disasters and other ecological threats. Despite steps in the right direction, national and international responses to this challenge remain limited, and protection for those affected inadequate. There is no clear definition of a 'climate refugee', nor are climate refugees covered by the 1951 Refugee Convention.

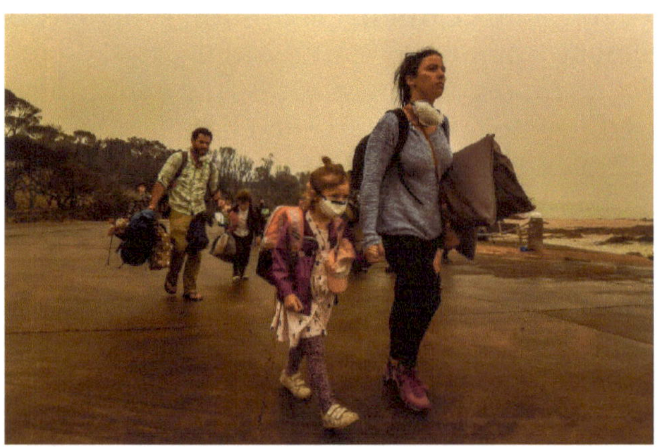

Climate refugees are people who must leave their homes and communities because of the effects of climate change and global warming. Climate refugees belong to a larger group of immigrants known as environmental refugees.

Source:https://legalaiddnlu.wordpress.com/wp-content/uploads/2020/06/rtx3i3ks-e1510158445731.jpg?w=1200
Reference link: https://education.nationalgeographic.org/reSource/environmental-refugee/

CLIMATE FRONTIER

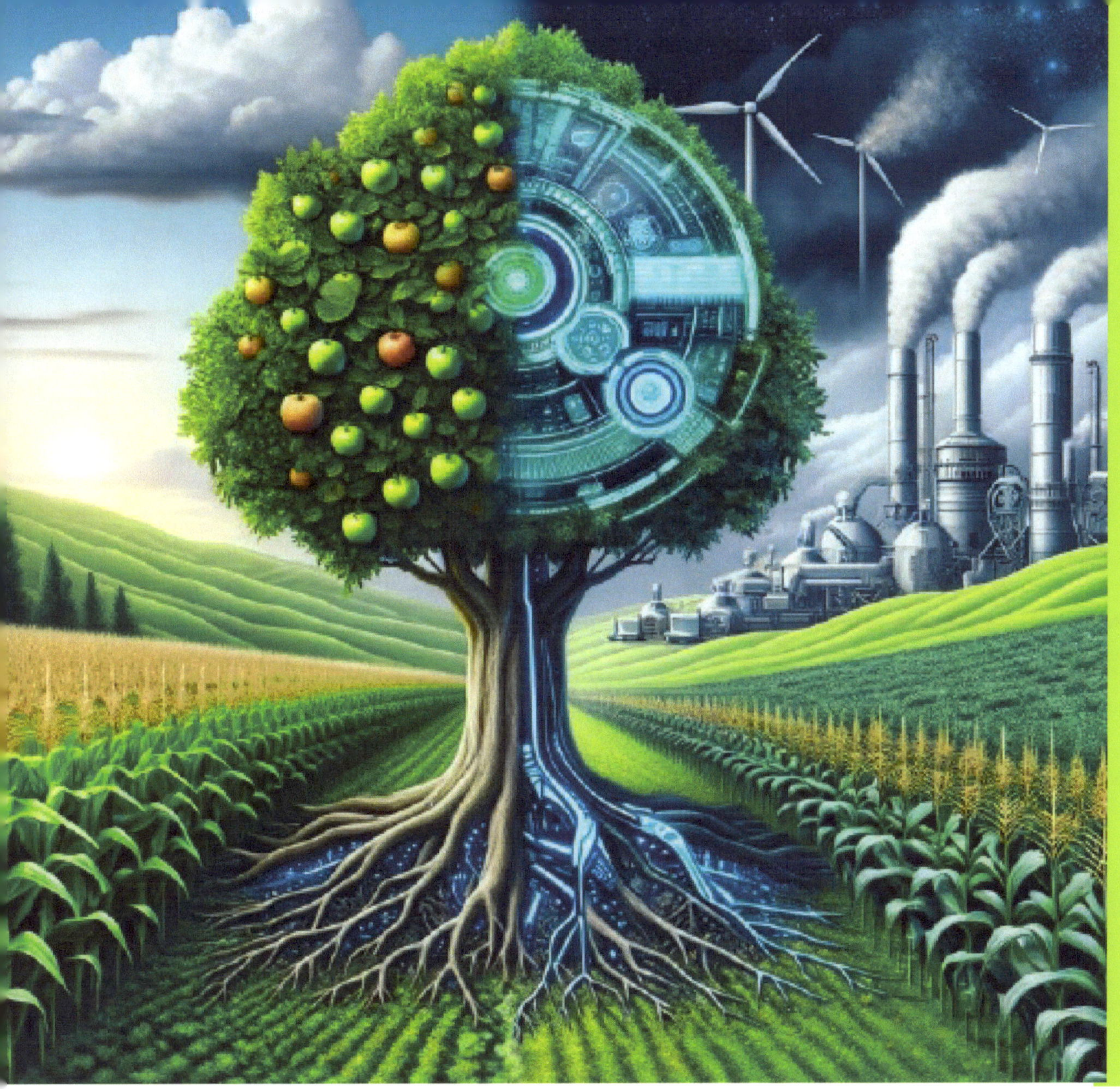

Climate frontiers represent the evolving and pioneering areas of addressing climate change and its impacts. This term covers various aspects of climate action and research. Scientifically, it includes advancements in understanding complex climate systems, improving climate models, and predicting future scenarios with greater accuracy. Technological innovations play a crucial role, involving the development of new renewable energy Sources, carbon capture and storage technologies, and climate-resilient infrastructure. On the policy front, climate frontiers encompass the formulation of novel regulatory frameworks, international agreements, and strategies to implement effective climate action at both national and global levels. These policies aim to mitigate greenhouse gas emissions, adapt to climate impacts, and support sustainable development goals. Additionally, addressing climate change at the community level involves promoting behavioral changes and sustainable practices among individuals and organizations. This includes encouraging energy conservation, waste reduction, and climate-conscious decision-making. Overall, exploring these frontiers is essential for creating integrated solutions to combat climate change. It requires interdisciplinary collaboration, innovation, and global cooperation to make significant progress and achieve long-term climate resilience.

Source: https://media.licdn.com/dms/image/D4E12AQHGle_xCskO0A/article-cover_image-shrink_720_1280/0/1713024613007?e=2147483647&v=beta&t=DLLlPr2BOEKrF-HoREq13rqlqoyS-_gfyE6d_OtGiOdE
Reference link: ChatGPT

CLIMATE DISCUSSIONS

Climate discussions encompass a wide range of topics related to the Earth's climate, including

CLIMATE CHANGE

The ongoing changes in global or regional climate patterns, particularly those related to the rise in global average temperatures due to human activities like fossil fuel burning, deforestation, and industrial processes.

GREENHOUSE GASES

These gases, such as carbon dioxide, methane, and nitrous oxide, trap heat in the Earth's atmosphere, contributing to global warming.

GLOBAL WARMING

The long-term increase in Earth's average surface temperature due to human activities. It is a key aspect of climate change and leads to various environmental impacts.

RENEWABLE ENERGY

The transition from fossil fuels to renewable energy Sources like solar, wind, and hydroelectric power is a critical component of mitigating climate change.

BIODIVERSITY

Climate change is affecting ecosystems and species worldwide, leading to shifts in habitats, migration patterns, and, in some cases, extinction.

CARBON FOOTPRINT

The total amount of greenhouse gases generated by our actions is a major topic in individual and corporate responsibility in the crisis.

ENVIRONMENTAL JUSTICE

Discussions often include the disproportionate impact of climate change on vulnerable populations, including low-income communities and developing countries.

CLIMATE ADAPTATION

As climate impacts become more severe, there is increasing focus on adapting infrastructure, agriculture, and societies to a changing climate.

TECHNOLOGICAL SOLUTIONS

Innovations such as carbon capture and storage, electric vehicles, and sustainable agriculture practices are central to addressing change.

These discussions are critical for understanding and addressing the challenges posed by climate change, with a focus on both mitigation (reducing the severity of climate change) and adaptation (adjusting to its impacts).

Source: https://img.freepik.com/premium-photo/climate-change-thoughtprovoking-discussion-climate-change_956369-9774.jpg   https://giki.earth/wp-content/uploads/2024/05/Blog-climate-conversations.png
Reference link: ChatGPT

CLIMATE POLICY

Climate Policy is a world-leading peer-reviewed academic journal, publishing high quality research and analysis on all aspects of climate policy, including mitigation and adaptation. Climate Policy aims to make high-quality research accessible and relevant, not only to academics, but also to policymakers and practitioners. The journal provides a platform for new ideas, innovative approaches and research-based insights that can help advance an effective response to climate change in practice.

CLIMATE POLICY COVERS THE FOLLOWING TOPICS:

Adaptation, mitigation, governance, and negotiations

Policy design, implementation and impact

Climate Policy is an interdisciplinary journal, and actively encourages submissions from all academic fields. A central requirement is that papers must be written in a language and style that avoids jargon, and is therefore accessible to academics from other fields as well as to policy practitioners and policymakers. Technical content can be included in online supplementary material.

Climate Policy Recommendations aerial photograph of harmful algal blooms over Lake Erie Climate change is one of the biggest threats to human communities and the long-term survival of America's wildlife. Impacts include worsening mega fires and hurricanes; harmful algal outbreaks; habitat loss; the spread of disease, pests, and invasive species; and a host of other dangerous conditions for people and wildlife. The climate science is clear—the world must slash greenhouse gas emissions in half by 2030, and reach net zero emissions by 2050. As such, we need practical policy solutions that will mitigate the impacts of climate change by quickly cutting and capturing carbon pollution and ensuring we are able to adapt to the impacts we cannot avoid.

Source: https://www.irisintelligence.com/wp-content/uploads/elementor/thumbs/ESG-Risks-and-Opportunities-psa2m0v10b44dih0r2efg9ts99w34wxzuhc1sfiu24.png
https://www.ft.com/__origami/service/image/v2/images/raw/https%3A%2F%2Fd1e00ek4ebabms.cloudfront.net%2Fproduction%2Fae731f0e-a842-4782-984b-3d7e001ee403.jpg?-source=next-article&fit=scale-down&quality=highest&width=700&dpr=1 Reference link:https://www.tandfonline.com/journals/tcpo20/about-this-journal#advertising-information

CLIMATE SCIENTISTS

Climate scientists measure changes to the planet using a variety of techniques. They study ice cores taken from the North and South poles to gauge changes to the environment over centuries. They sample air to see how concentrations of greenhouse gases are evolving and examine trees to chart how forest conditions have changed over time. They build computer models that predict changes to ecosystems based on historical data. These and other methods illuminate what causes changes to the climate, what effects those changes have on ecosystems, how they will affect wildlife and humans, what can be done to slow or reverse harmful climate trends, and how humans and animals might adapt to a changing world.

Disciplines related to climatology include atmospheric science, hydrology, environmental microbiology, cloud physics, meteorology, and Earth system modeling. Scientists have been studying Earth's climate for hundreds of years. One of the earliest discoveries came from Jan Baptista van Helmont, a Flemish chemist who in the 1640s established the idea of gases and coined the name "gas" from the Greek word for chaos. Among his scientific contributions, van Helmont discovered a gas that would later become known as carbon dioxide. In the 1800s, scientists began understanding the role of gases in trapping heat in the atmosphere. The French mathematician Joseph Fourier first described the idea that Earth's atmosphere retains the sun's heat in 1824. In the 1850s, American scientist Eunice Foote and British physicist John Tyndall independently documented the heat-trapping capabilities of carbon dioxide and water vapour. And in the late 1890s, Swedish chemist Svante Arrhenius calculated how changes in carbon dioxide in the air would affect surface temperatures.

Source: https://media.springernature.com/full/springer-static/image/art%3A10.1 038%2F525449a/MediaObjects/41586_2015_Article_BF525449a_Figa_HTML.jpg
Reference link: https://www.pnnl.gov/explainer-articles/climate-science#:~:text=Climate%20scientists%20measure%20changes%20to,to%20the%20environment%20over%20centuries

CLIMATE GOAL

A climate target, climate goal or climates pledge is a measurable long-term commitment for climate policy and energy policy with the aim of limiting the climate change. Researchers within, among others, the UN climate panel have identified probable consequences of global warming for people and nature at different levels of warming. Based on this, politicians in a large number of countries have agreed on temperature targets for warming, which is the basis for scientifically calculated carbon budgets and ways to achieve these targets. This in turn forms the basis for politically decided global and national emission targets for greenhouse gases, targets for fossil-free energy production and efficient energy use, and for the extent of planned measures for climate change mitigation and adaptation.

Targets are the limits that scientists and policymakers set in plans to combat climate change. These targets can take different forms, from goals for limiting the Earth's warming to hard caps on greenhouse gas emissions. For example, the Paris Agreement set 2 as a temperature target for global warming, while the state of Massachusetts has a target to reduce emissions to 50% below 1990 levels by 2030.

Source: https://i0.wp.com/epthinktank.eu/wp-content/uploads/2020/12/eprs-briefing-659370-climate-target-plan-final.jpeg.png?fit=820%2C486&ssl=1
Reference link https://en.wikipedia.org/wiki/Climate_target#:~:text=A%20climate%20target%2C%20climate%20goal,of%20limiting%20the%20climate%20change
https://climate.mit.edu/explainers/climate-targets#:~:text=These%20targets%20can%20take%20different,below%201990%20levels%20by%202030

CLIMATE TRANSPARENCY

Climate Transparency is a global partnership with a shared mission to stimulate a "race to the top" in climate action in G20 countries through enhanced transparency.

Joint assessment to increase transparency

Climate Transparency brings together the most authoritative climate assessments and expertise of stakeholders from G20 countries. Jointly, these experts develop a credible, comprehensive and comparable picture on G20 climate performance: The Climate Transparency Report covers easy-to-use information on all major areas such as mitigation and climate finance and includes detailed fact sheets on all G20 countries. It is published on an annual basis on the eve of the G20 Summit.

Empowering change agents

Climate Transparency aims to increase awareness and peer pressure among policy-makers in G20 governments and influencers from civil society and the financial sector on a national and international level. Activities of Climate Transparency to empower change agents include international and country-specific media work, direct communication with decision makers and G20 engagement groups as well as workshops in G20 countries. Climate Transparency was co-founded under the leadership of Peter Eigen (Founder of Transparency International) and Alvaro Umana (former Minister of Environment and Energy of Costa Rica) in late 2014, who also serve as Co-Chairs of the Partnership to this day. Climate Transparency is made possible through support from the Federal Ministry for the Environment, Nature Conservation and Nuclear Safety (BMU) through the International Climate Initiative, the Climate Works Foundation, the World Bank Group and the Federal Foreign Office.

Source:https://www.giz.de/static/en/images/ContentImages_460x160px_ELS/Article%20Thumbnail-Banner_rdax_658x370.png
Reference link: https://www.climate-transparency.org/about-us

CLIMATE ECONOMICS

Climate economics is the study of the economic impact of climate change, and how to use economic models and analysis to guide policies and adaptation strategies. "Economic analyses play a critical role in consideration of climate change policies. Identifying, assessing and communicating the implications of economic uncertainty and knowledge gaps remains a major challenge - for example - in characterization of long-term technology change and valuation of non-market impacts. I believe that Climate Change Economics can provide an important forum to consider fundamental economic issues that will enhance understanding and improve climate policy deliberations." — Dr. Brian P. Flannery, Science, Strategy and Programs Manager, Exxon Mobil Corporation.

An economic analysis of climate change uses economic tools and models to calculate the magnitude and distribution of damages caused by climate change. It can also give guidance for the best policies for mitigation and adaptation to climate change from an economic perspective. There are many economic models and frameworks. For example, in a cost–benefit analysis, the trade offs between climate change impacts, adaptation, and mitigation are made explicit. For this kind of analysis, integrated assessment models (IAMs) are useful. Those models link main features of society and economy with the biosphere and atmosphere into one modelling framework.The total economic impacts from climate change are difficult to estimate.

In general, they increase the more the global surface temperature increases Many effects of climate change are linked to market transactions and therefore directly affect metrics like GDP or inflation. However, there are also non-market impacts which are harder to translate into economic costs. These include the impacts of climate change on human health, biomes and ecosystem services. Economic analysis of climate change is challenging as climate change is a long-term problem. Furthermore, there is still a lot of uncertainty about the exact impacts of climate change and the associated damages to be expected. Future policy responses and socioeconomic development are also uncertain.

Economic analysis also looks at the economics of climate change mitigation and the cost of climate adaptation. Mitigation costs will vary according to how and when emissions are cut. Early, well-planned action will minimize the costs. Globally, the benefits of keeping warming under 2 °C exceed the costs.Cost estimates for mitigation for specific regions depend on the quantity of emissions allowed for that region in future, as well as the timing of interventions. Economists estimate the cost of climate change mitigation at between 1% and 2% of GDP. The costs of planning, preparing for, facilitating and implementing adaptation are also difficult to estimate, depending on different factors. Across all developing countries, they have been estimated to be about USD 215 billion per year up to 2030, and are expected to be higher in the following years.

Source:https://www.worldscientific.com/worldscinet/cce?srsltid=AfmBOormN5bCWYR64U3Z6mR87XMJAVgixXWaEsHgloTY-qLKmY7odRnY https://en.wikipedia.org/wiki/Economic_analysis_of_climate_change#:~:text=Economic%20analysis%20of%20climate%20change%20is%20an%20umbrella%20term%20for,preventing%20or%20softening%20those%20effects.

CLIMATE LEADERSHIP

Climate leadership refers to the actions and decisions made by individuals, organizations, and governments to reduce greenhouse gas emissions, mitigate the impacts of climate change, and promote sustainable development.

Effective climate leadership involves:
1. Setting ambitious emission reduction targets
2. Developing and implementing climate-resilient policies
3. Investing in renewable energy and clean technologies
4. Promoting sustainable land use and forestry practices
5. Supporting climate change research, development, and deployment
6. Encouraging climate education, awareness, and community engagement
7. Fostering international cooperation and climate diplomacy

Examples of climate leadership include:
1. Countries like Norway and Costa Rica, which have made significant commitments to renewable energy and sustainable development
2. Cities like Copenhagen and Vancouver, which have set ambitious emission reduction targets and implemented innovative climate solutions
3. Companies like Patagonia and IKEA, which have integrated sustainability and climate action into their business models
4. Individuals like Greta Thunberg and David Attenborough, who have raised awareness about the urgent need for climate action

Individual Actions
1. Reduce energy consumption: Use energy-efficient appliances, turn off lights, and insulate homes.
2. Switch to renewable energy: Invest in solar panels or renewable energy credits.
3. Eat a plant-based diet: Animal agriculture contributes to greenhouse gas emissions.
4. Conserve water: Take shorter showers and fix leaks.
5. Reduce, reuse, recycle: Minimize single-use plastics and recycle.

Community Involvement
1. Participate in climate protests and events: Raise awareness and demand action from leaders.
2. Join a local climate group: Collaborate with others to promote sustainability.
3. Advocate for climate policies: Contact representatives and support climate-friendly legislation.

CLIMATE POLITICS

The politics of climate change results from different perspectives on how to respond to climate change. Global warming is driven largely by the emissions of greenhouse gases due to human economic activity, especially the burning of fossil fuels, certain industries like cement and steel production, and land use for agriculture and forestry. Since the Industrial Revolution, fossil fuels have provided the main Source of energy for economic and technological development. The centrality of fossil fuels and other carbon-intensive industries has resulted in much resistance to climate friendly policy, despite widespread scientific consensus that such policy is necessary. Climate change first emerged as a political issue in the 1970s.

Efforts to mitigate climate change have been prominent on the international political agenda since the 1990s, and are also increasingly addressed at national and local level. Climate change is a complex global problem. Greenhouse gas (GHG) emissions contribute to global warming across the world, regardless of where the emissions originate. Yet the impact of global warming varies widely depending on how vulnerable a location or economy is to its effects. Global warming is on the whole having negative impact, which is predicted to worsen as heating increases. Ability to benefit from both fossil fuels and renewable energy Sources vary substantially from nation to nation. Different responsibilities, benefits and climate related threats faced by the world's nations contributed to early climate change conferences producing little beyond general statements of intent to address the problem, and non-binding commitments from the developed countries to reduce emissions. In the 21st century, there has been increased attention to mechanisms like climate finance in order for vulnerable nations to adapt to climate change. In some nations and local jurisdictions, climate friendly policies have been adopted that go well beyond what was committed to at international level. Yet local reductions in GHG emission that such policies achieve have limited ability to slow global warming unless the overall volume of GHG emission declines across the planet.

CLIMATE LEGISLATION

Climate change legislation, often abbreviated to climate legislation, consists of the laws and policies that govern action on climate change by setting its legal basis. The term is used broadly and can include acts, decrees and policies that are passed or promulgated by both legislative and executive branches of government. These laws and policies address actions that fall under the scope of climate change mitigation, adaptation and disaster risk management. Laws and policies can apply across different sectors or focus on one, such as agriculture, land use, transport, energy, waste, environment, tourism, industry, buildings, water and health.

The term climate 'framework legislation' describes an important subset of climate laws that link national and international agendas. Framework climate laws share some or all of the following characteristics: they set out the strategic direction for national climate change policy; are passed by the legislative branch of government; contain national, long-term and/or medium targets and/or pathways for change; set out institutional arrangements for climate governance at the national level; are multi-sectoral in scope; and involve mechanisms for transparency and/or accountability.

HOW DOES DOMESTIC CLIMATE LEGISLATION RELATE TO INTERNATIONAL CLIMATE CHANGE LAW?

World leaders first expressly recognized that climate change needed to be tackled with institutions, rules and procedures at the Rio Earth Summit in 1992. The United Nations Framework Convention on Climate Change (UNFCCC) was subsequently established, with the primary goal of promoting international action to stabilize greenhouse gas emissions. The convening of the UNFCCC parties has led to several milestones, most notably the 2015 Paris Agreement. The Agreement covers mitigation, adaptation and financial burden-sharing, and relies on pledges (the so-called Nationally Determined Contributions – NDCs) made by each UNFCCC party to achieve its goal of limiting global warming to well below 2°C, preferably 1.5°C, above pre-industrial levels.

CLIMATE STRESS TESTING

Climate stress tests are frequently mentioned as a key element in central banks' tool-kit to address climate-related risks to the financial system. Several countries have exercises in the works, although some analysts say they fall short of the comprehensive climate stress tests that are required, and it could be many years before their benefits are felt.

STRESS TESTS EMERGED AFTER THE FINANCIAL CRASH AND HAVE SINCE BEEN APPLIED TO CLIMATE

Stress tests, led by prudential regulators, are assessments of how well banks are able to cope with financial and economic shocks. They allow supervisors to identify banks' vulnerabilities and work with those institutions to address them. Stress testing was one of the measures institutionalised by the Basel Accords after the 2008 financial crisis to decrease damage to the economy resulting from banks taking on too much risk.

Source: https://www.enterpriseitworld.com/wp-content/uploads/2023/02/Stress-Testing1.jpg Reference link: https://greencentralbanking.com/2022/03/14/what-are-climate-stress-tests/

CLIMATE SHOCKS/ CLIMATE CHAOS

CLIMATE SHOCKS

Climate shocks are sudden, unexpected events that significantly disrupt the climate system or the environment. These shocks often have immediate and severe impacts on ecosystems, economies, and human communities. Examples include

Climate shocks can lead to short-term and long-term challenges, such as disruptions in food and water supply, health crises, and economic losses.

CLIMATE CHAOS

Climate chaos refers to the broader and more complex impacts of climate change that result in a high degree of unpredictability and disorder in weather patterns and environmental conditions. This term encompasses

Climate chaos reflects the broader, more systemic impacts of climate change, characterized by increasing unpredictability and instability in climate-related conditions, which can exacerbate vulnerabilities and complicate adaptation and mitigation efforts.

Source: https://scitechdaily.com/images/Bad-Climate-Change-Concept.jpg
Reference link: ChatGPT

GREEN HOUSE GASES (GHG)

Greenhouse gases (GHGs) are the gases in the atmosphere that raise the surface temperature of planets such as the Earth. What distinguishes them from other gases is that they absorb the wavelengths of radiation that a planet emits, resulting in the greenhouse effect. The Earth is warmed by sunlight, causing its surface to radiate heat, which is then mostly absorbed by greenhouse gases. Without greenhouse gases in the atmosphere, the average temperature of Earth's surface would be about −18 °C (0 °F), rather than the present average of 15 °C (59 °F).

The five most abundant greenhouse gases in Earth's atmosphere, listed in decreasing order of average global mole fraction, are water vapor, carbon dioxide, methane, nitrous oxide, ozone. Other greenhouse gases of concern include chlorofluorocarbons hydro fluorocarbons (HFCs), per fluorocarbons, SF and NF. Water vapor causes about half of the greenhouse effect, acting in response to other gases as a climate change feedback.

Human activities since the beginning of the Industrial Revolution (around 1750) have in-creased carbon dioxide by over 50%, and methane levels by 150%. Carbon dioxide emissions are causing about three-quarters of global warming, while methane emissions cause most of the rest. The vast majority of carbon dioxide emissions by humans come from the burning of fossil fuels, with remaining contributions from agriculture and industry. Methane emissions originate from agriculture, fossil fuel production, waste, and other Sources. The carbon cycle takes thousands of years to fully absorb CO_2 from the atmosphere, while methane lasts in the atmosphere for an average of only 12 years.

Natural flows of carbon happen between the atmosphere, terrestrial ecosystems, the ocean, and sediments. These flows have been fairly balanced over the past 1 million years, although greenhouse gas levels have varied widely in the more distant past. Carbon dioxide levels are now higher than they have been for 3 million years. If current emission rates continue then global warming will surpass 2.0 °C (3.6 °F) sometime between 2040 and 2070. This is a level which the Intergovernmental Panel on Climate Change (IPCC) says is "dangerous".

Source: https://www.alpinme.com/files/expo2020-1-3200-x-1800-1-1.jpg Reference link: https://en.wikipedia.org/wiki/Greenhouse_gas

GHG EMISSIONS

Greenhouse gas emissions are gases that trap heat in the atmosphere, contributing to global warming and climate change. The most significant greenhouse gases include

CARBON DIOXIDE (CO$_2$)

SOURCE	Combustion of fossil fuels (coal, oil, and natural gas), deforestation, cement production, and certain industrial processes.
ROLE	CO$_2$ is the most significant greenhouse gas released by human activities, contributing substantially to global warming.

METHANE (CH$_2$)

SOURCE	Agricultural practices (especially livestock digestion), landfills, wetlands, and natural gas production and distribution.
ROLE	Methane is more effective than CO at trapping heat, although it is present in smaller quantities in the atmosphere. Its warming potential is significantly higher than CO over a short time frame.

NITROUS OXIDE (N$_2$O)

SOURCE	Agricultural activities (use of synthetic fertilizers), industrial processes, and fossil fuel combustion.
ROLE	Nitrous oxide has a strong greenhouse effect and also contributes to ozone depletion in the stratosphere.

CHLOROFLUOROCARBONS (CFCS) AND OTHER FLUORINATED GASES

SOURCE	Industrial applications, refrigerants, air conditioning, and certain cleaning products.
ROLE	These gases have a high global warming potential and long atmospheric lifetimes.

Source:https://upload.wikimedia.org/wikipedia/commons/3/34/Reduce_greenhouse_gas_emissions.png
Reference link: ChatGPT

GREEN TECHNOLOGY

Environmental technology is the use of engineering and technological approaches to understand and address issues that affect the environment with the aim of fostering environmental improvement. Green tech is a type of technology that is considered environmentally friendly based on its production process or its supply chain. The term "green technology" or "green tech" can also apply to technologies that produce clean energy, utilize alternative fuels, and are less harmful to the environment than fossil fuels. Green technologies are designed to protect ecosystems and endangered species, reduce carbon emissions, and conserve natural re-Sources. Examples of green tech include solar power, electric vehicles, sustainable agriculture, and carbon capture.

There is no commonly accepted or internationally agreed definition of green technology. The term can be broadly defined as technology that has the potential to significantly improve environmental performance relative to other technology. It is related to the term "environmentally sound technology", which was adopted under the United Nations Conference on Environment and Development Agenda 21, although it is no longer widely used. Based on Agenda 21, environmentally sound technologies are geared to "protect the environment, are less polluting, use all resources in a more sustainable manner, recycle more of their wastes and products, and handle residual wastes in a more acceptable manner than the technologies for which they were substituted."1 Other related terms for green technology include climate-smart, climate-friendly and low-carbon technology.

GREEN ECONOMY

A green economy is an economy that aims at reducing environmental risks and ecological scarcities, and that aims for sustainable development without degrading the environment. It is closely related with ecological economics, but has a more politically applied focus. The 2011 UNEP Green Economy Report argues "that to be green, an economy must not only be efficient, but also fair. Fairness implies recognizing global and country level equity dimensions, particularly in assuring a Just Transition to an economy that is low-carbon, reSource efficient, and socially inclusive."

 The term green economy was first coined in a pioneering 1989 report for the Government of the United Kingdom by a group of leading environmental economists, entitled Blueprint for a Green Economy (Pearce, Markandya and Barbier, 1989). The report was commissioned to advise the UK Government if there was a consensus definition to the term "sustainable development" and the implications of sustainable development for the measurement of economic progress and the appraisal of projects and policies. Apart from in the title of the report, there is no further reference to green economy and it appears that the term was used as an afterthought by the authors. In 1991 and 1994 the authors released sequels to the first report entitled Blueprint 2 Greening the world economy and Blueprint 3 Measuring Sustainable Development. Whilst the theme of the first Blueprint report was that economics can and should come to the aid of environmental policy, the sequels extended this message to the problems of the global economy - climate change, ozone depletion, tropical deforestation, and reSource loss in the developing world. All reports built upon research and practice in environmental economics spanning back several decades.

Source:https://miro.medium.com/v2/resize:fit:1400/1*RoQDNvivJs3M5_HkDPCUbg.png https://miro.medium.com/v2/resize:fit:1400/1*RoQDNvivJs3M5_HkDPCUbg.png
Reference link: https://en.wikipedia.org/wiki/Green_economy https://sustainabledevelopment.un.org/index.php?menu=1446

GREEN CREDIT

Green Credit refers to a unit of incentive provided to individuals and entities engaged in activities that deliver a positive impact on the environment. It is a voluntary program initiated by the government to incentivize various stakeholders in contributing to environmental preservation and sustainable practices. This program is part of the broader 'LiFE' campaign (Lifestyle for Environment), and it encourages and rewards voluntary environmentally-positive actions.

COVERED ACTIVITIES

The Green Credit program encompasses eight key types of activities aimed at enhancing environmental sustainability

TREE PLANTATION	- Planting trees to increase green cover and combat deforestation.
WATER MANAGEMENT	- Implementing strategies to efficiently manage and conserve water resources.
SUSTAINABLE AGRICULTURE	- Promoting eco-friendly and sustainable agricultural practices.
WASTE MANAGEMENT	- Implementing effective waste management systems to reduce environmental pollution.
AIR POLLUTION REDUCTION	- Initiatives aimed at reducing air pollution and improving air quality.
MANGROVE CONSERVATION AND RESTORATION	- Protecting and restoring mangrove ecosystems for ecological balance.

GREEN WASHING

Green washing refers to the act of making false or misleading claims about the positive environmental impact that a company, product or service has on the environment. Green washing occurs when organizations present untrue actions or statements that appear more environmentally friendly than they actually are. The term Green washing was first coined in 1986 by environmentalist Jay Westerveld in an article where he decried the common practice of hotels asking guests to reuse towels to help conserve energy. Westerveld claimed that those same hotels did little to help the environment and that the towel request was an act of Green washing.

SIX SHADES OF GREEN WASHING

1. Green Crowding - This is clever. It relies on safety in numbers and happens when different groups (like governments, organizations and companies) join forces to effect change

2. Green lighting – It is when companies spotlight a particularly 'green' product or operation leaving everything else in the shadows.

3. Green shifting – It shifts the blame onto consumers. BP's "Know your carbon footprint" campaign is a key example.

4. Green labeling – It happens when companies use words like 'eco', 'sustainable' or related wording or symbols conveying green messaging with no evidence to support it.

5. Green rinsing – It is where companies change their sustainability commitments or targets before achieving them.

6. Green hushing – It happens when companies deliberately underreport or hide green credentials to evade scrutiny.

Source: https://www.ippi.org.il/wp-content/uploads/2023/03/Greenwashing_image.png' Reference link: https://www.techtarget.com/whatis/definition/greenwashing#:~ https://enviral.co.uk/the-new-shades-of-greenwash/

GREEN CREDIT AUDITORS

Green credit auditors are professionals who evaluate projects to ensure compliance with environmental sustainability standards, often tied to carbon credits or green financing. Their primary role is to assess whether projects meet the requirements for receiving "green credits," which represent efforts to reduce greenhouse gas emissions or promote sustainable practices. These credits can be traded or sold, helping businesses and organizations offset their carbon footprint and contribute to global climate goals.

Auditors conduct thorough reviews of project documentation, analyze data related to environmental impact, and perform site visits to validate claims made by the project developers. They verify that the project complies with specific environmental frameworks, such as carbon reduction targets, renewable energy usage, or waste management practices.

Green credit auditors work with internationally recognized standards, such as the Gold Standard, Verified Carbon Standard (VCS), or Green-e®. Their evaluations are essential for certifying the environmental integrity of projects, allowing businesses to claim green credits in the carbon markets. By ensuring transparency and accountability, these auditors play a critical role in promoting sustainable practices and facilitating the transition to a low-carbon economy.

Source: https://www.shutterstock.com/shutterstock/photos/565316725/display_1500/stock-vector-financial-audit-circular-illustration-vector-colorful-auditing-round-creative-symbol-565316725.jpg https://www.shutterstock.com/shutterstock/photos/565316725/display_1500/stock-vector-financial-audit-circular-illustration-vector-colorful-auditing-round-creative-symbol-565316725.jpg Reference link: ChatGPT

GREEN CREDIT VERIFIER

A green credit verifier plays a crucial role in ensuring that environmental projects meet the criteria for earning green credits. These credits represent a quantifiable reduction in greenhouse gas emissions or other sustainability efforts, which can be traded or used to offset a company's carbon footprint. The verifier acts as an independent party, reviewing projects to validate that they comply with established standards and genuinely contribute to environmental goals.

The verification process involves a comprehensive review of project documentation, data, and, in many cases, a physical inspection of the project site. The verifier assesses whether the project has accurately measured its emissions reductions or sustainability improvements. They ensure the methodology aligns with international standards, such as those set by *Gold Standard* or *Verified Carbon Standard (VCS)

For example, a project aiming to reduce carbon emissions through reforestation might submit data showing how many trees have been planted and how much carbon they are expected to absorb over time. A green credit verifier would confirm these figures, review the data collection methods, and ensure the project adheres to strict environmental guidelines. Only after passing this verification can the project issue carbon credits that can be sold or traded. Green credit verifiers work with well-established programs, such as the *Green-e® Climate* certification, which certifies projects based on their environmental impact and transparency. They ensure that projects are legitimate and that the credits issued are based on real, measurable environmental benefits.

By ensuring transparency, accuracy, and adherence to best practices, green credit verifiers help build trust in the carbon market. Their work is essential for promoting responsible investment in sustainability projects and ensuring that organizations can confidently use carbon credits to meet their environmental targets. Verifiers provide the critical oversight needed to ensure that green credits genuinely contribute to reducing global emissions and promoting a more sustainable future.

Source: https://thepaypers.com/Images/environment-sustainability.jpg
Reference link: ChatGPT

GREEN CREDIT TRADING PLATFORM

Agreen credit trading platform is an online marketplace that facilitates the buying and selling of green credits, such as carbon credits, which represent measurable reductions in greenhouse gas emissions. These platforms enable businesses, governments, and individuals to purchase credits to offset their carbon footprint or meet regulatory compliance, while project developers can sell credits generated from sustainable initiatives like renewable energy, reforestation, or energy efficiency improvements.

The trading platform acts as a bridge between credit buyers and sellers, ensuring transparency and accountability in the transactions. Credits listed on the platform are typically verified by independent auditors to ensure they meet recognized environmental standards, such as those set by Verified Carbon Standard (VCS) or Gold Standard. After verification, the credits are recorded in a green credit registry to prevent double-counting or fraudulent claims.

An example of a widely used green credit trading platform is Climate Trade, which allows businesses to offset their carbon emissions by purchasing verified carbon credits. Another example is Air Carbon Exchange a global marketplace for carbon credits that provides real-time pricing, transparency, and liquidity for carbon trading.

These platforms help scale climate action by making it easier for companies to invest in sustainability and reach their emissions reduction targets. By providing a transparent and secure marketplace, green credit trading platforms promote a more efficient carbon market and contribute to the global effort to combat climate change.

Source: https://image.chitra.live/api/v1/wps/758459f/c41acc9a-af4d-4c72-a22c-32aaefa88216/6/iStock-1388420740-679x419.jpg
Reference link: ChatGPT

GREEN CREDIT REGISTRY

A green credit registry is a platform that records and tracks the issuance, ownership, and retirement of green credits, such as carbon credits. These registries ensure transparency, accountability, and credibility in the carbon markets by providing a publicly accessible record of transactions. Green credits represent reductions in greenhouse gas emissions, which can be traded or used to offset an organization's carbon footprint.

The registry functions as a centralized system where projects that generate green credits (like reforestation or renewable energy projects) are listed and their credits are verified by independent auditors before being issued. Once verified, the credits are recorded in the registry and can be sold or retired (when used to offset emissions). This prevents double-counting or fraudulent claims of emissions reductions.

Prominent green credit registries include the Verified Carbon Standard (VCS) Registry managed by Verra, which is widely used in voluntary carbon markets. Similarly, the Gold Standard Registry is another trusted platform for carbon credits, emphasizing sustainable development goals.

By offering a transparent record, green credit registries provide confidence to buyers and sellers that the credits represent real, verified environmental benefits, helping to drive investment in sustainable projects.

GREEN FUTURE

A green future refers to a vision for a more environmentally sustainable and ecologically-friendly society. Key elements of a green future typically include:

- Transitioning to renewable energy Sources like solar, wind, and hydroelectric power to reduce fossil fuel use and greenhouse gas emissions.

- Promoting energy efficiency in buildings, transportation, and industry to minimize energy consumption.

- Shifting towards a more circular economy that minimizes waste, reuses/recycles materials, and relies on renewable and biodegradable resources.

- Protecting and restoring natural ecosystems like forests, wetlands, and oceans to preserve biodiversity and the vital functions they provide.

- Developing sustainable agricultural practices that work in harmony with the environment rather than degrading it.

- Implementing policies and infrastructure to support sustainable modes of transportation like walking, cycling, and public transit.

- Investing in research and technologies that can enable a more environmentally-friendly way of living and doing business.

The overarching goal of a green future is to live within the means of the planet's natural resources and ecosystems, meeting present needs without compromising the ability of future generations to do the same. It represents a transition to a more sustainable, resilient, and environmentally-conscious society.

Source https://akm-img-a-in.tosshub.com/businesstoday/images/story/201608/green660_082216010208.jpg Reference link: https://www.quora.com/What-is-a-green-future

GREEN MANUFACTURING

Green manufacturing is a commitment to using fewer environmental pollutants and natural resources. Green facilities also make every effort to produce less waste, carbon emissions, and try to have as little impact on the environment as possible. 'Eco-friendly' initiatives are allowing many businesses to reduce costs and run a more productive operation. Additionally, many organizations are making the process of becoming sustainable easier than ever. Below, we outline how manufacturers are becoming more eco-friendly and how they are benefiting from the switch.

HOW ARE BUSINESSES GOING GREEN?

ALTERNATIVE ENERGY	An energy audit is a quick and easy way to start creating an eco-conscious facility. This audit takes into account all energy consumption in order to find areas of waste. After auditing energy use, businesses can determine the best way to correct these losses. Whether it be through energy-efficient lighting or by replacing outdated motors, these small changes can have a big payoff.
REDUCE WASTE	Businesses can also go through a waste audit to ensure they're not disposing of valuable materials. Some raw materials may be recycled or reduced for different projects. Over time, reducing a facility's overall waste will save money and increase efficiency.
USING NON-TOXIC MATERIALS	In addition to reducing waste and conserving energy, many companies are opting to use eco-friendly supplies. These brands are replacing synthetic materials with cleaner, safer ones. For example, traditional mattress brands use foam made of 100 percent petroleum-based oils. However, to reduce their carbon footprint, environmentally conscious mattresses replace a portion of the petroleum in the bed with plant-based oils. This results in less pollution and ensures the product is safe for consumers.

GREEN TRANSPORT

Green transportation is any means of travel that doesn't negatively impact the environment. Green transportation can be private (like a fast ebike), or public transit (like an electric city bus). Walking is also considered a green mode of travel. The common denominator among green transportation everywhere is that it's sustainable. Sustainable transportation is powered by resources that aren't depleted when used, making them harness able by future generations.

TYPES OF SUSTAINABLE (GREEN) VEHICLES

Looking to make a move toward green transportation? Here are three sustainable options people are embracing in 2021.

All-electric vehicles use one or more electric motors. They can be vehicles powered by solar cells, electric batteries, or electric generators to convert fuel to electricity. All-electric vehicles are greener than vehicles powered by combustion engines because they don't burn up fossil fuel(which isn't renewable), and they don't negatively impact the environment in the same way exhaust does either.

1

Hybrid vehicles operate on the premise that each power Source will do what it does best. Although hybrid vehicles are less "green" than all-electric vehicles, they still produce fewer greenhouse emissions than vehicles running purely on a combustion engine, making them greener. Depending on how they are driven, hybrid cars cut emissions from 26 to 90 percent when compared to traditional vehicles. Like all-electric vehicles, hybrid vehicles can be found in both private and public transportation.

2

Electric bikes are effective modes of green transportation for a number of reasons. For instance, ebike production requires fewer resources than full-sized vehicles, and they demand less energy too. E-bikes are just starting to become popular in the United States. However, they're already an increasingly popular form of green transportation in Europe and Asia. Especially in cities, where getting around the landscape on an ebike is easier for couriers and commuters.

3

Source: https://encrypted-tbn0.gstatic.com/images?q=tbn:ANd9GcQu-l3TWNy9U53EmyGjilKDzrTFVPc_L9xS4A&s https://www.trukky.com/blog/wp-content/uploads/2016/10/Green-Truck-Transportation-1.jpg Reference link: https://keegomobility.com/blog/what-is-green-transportation/

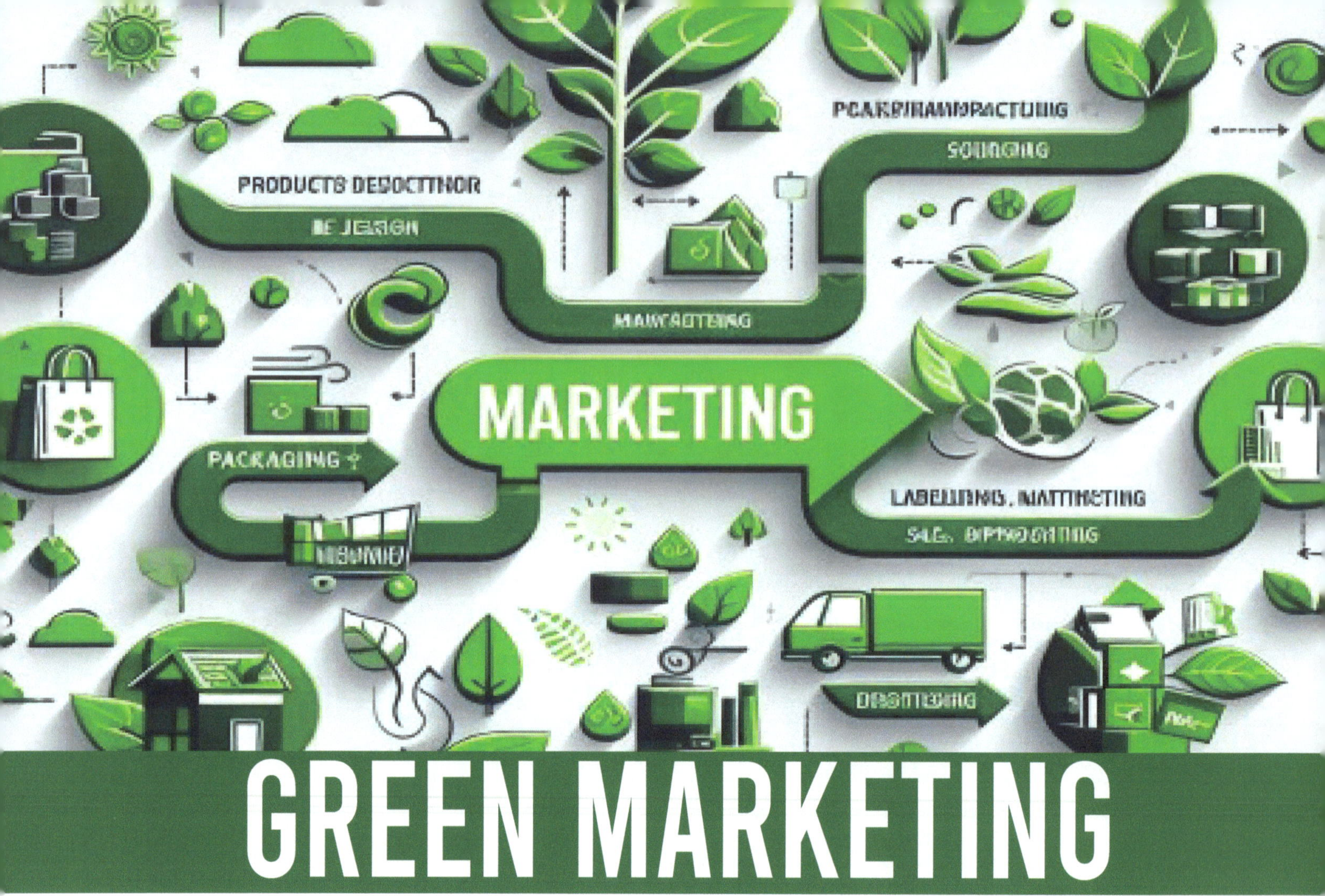

GREEN MARKETING

Green marketing refers to the practice of developing and advertising products based on their real or perceived environmental sustainability. Examples of green marketing include advertising the reduced emissions associated with a product's manufacturing process, or the use of post-consumer recycled materials for a product's packaging. Some companies also may market themselves as being environmentally-conscious companies by donating a portion of their sales proceeds to environmental initiatives, such as tree planting.

Green marketing describes a company's efforts to advertise the environmental sustainability of its business practices.The emergence of a consumer population that is becoming increasingly concerned with environmental and social factors has led to green marketing becoming an important component of corporate public relations.

 One criticism of green marketing practices is that they tend to favor large corporations that can absorb the additional costs entailed by these programs.Smaller businesses may not be able to shoulder the high-cost burden of green marketing, but this isn't to say, they cannot.
Green washing occurs when a company states it is involved in environmental endeavors but it turns out the claims can't be substantiated.

Source:https://timesproweb-static-backend-prod.s3.ap-south-1.amazonaws.com/Blog_Page_On_What_Is_Green_Marketing_d951998f94.webp
Reference link: https://www.investopedia.com/terms/g/green-marketing.asp

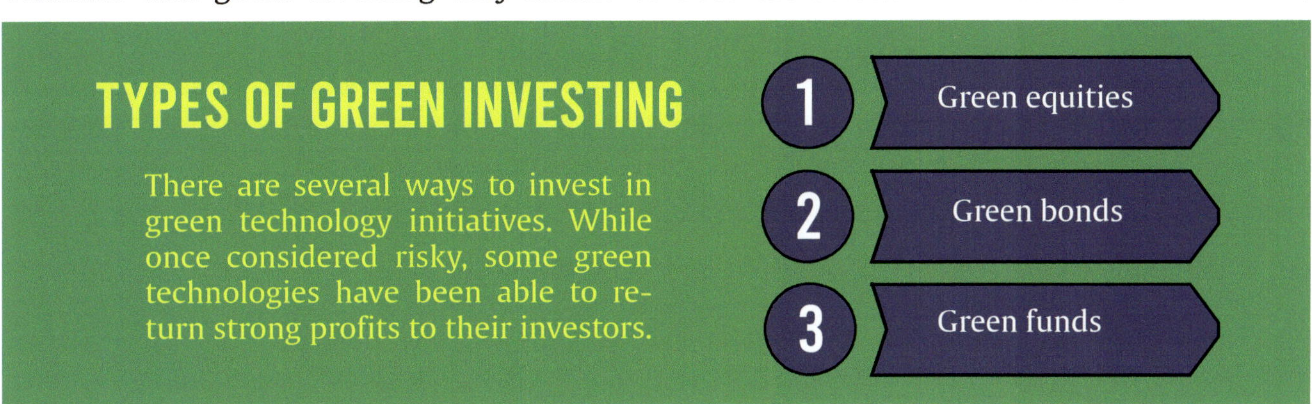

GREEN INVESTING

Green investing seeks to support business practices that have a favorable impact on the natural environment. Often grouped with socially responsible investing (SRI) or environmental, social, and governance (ESG) criteria, green investments focus on companies or projects committed to the conservation of natural reSources, pollution reduction, or other environmentally conscious business practices. Green investments may fit under the umbrella of SRI but are more specific.

Some investors buy green bonds, green exchange-traded funds (ETFs), green index funds, green mutual funds, or hold stock in environmentally friendly companies to support green initiatives. While profit is not the only motive for those investors, there is some evidence that green investing may mimic or beat the returns of more traditional assets.

TYPES OF GREEN INVESTING

There are several ways to invest in green technology initiatives. While once considered risky, some green technologies have been able to return strong profits to their investors.

1 Green equities

2 Green bonds

3 Green funds

Source: https://m.economictimes.com/thumb/msid-96470994,width-1200,height-900,resizemode-4,imgsize-77034/sustainable-investing-to-surge.jpg
Reference link: https://www.investopedia.com/terms/g/green-investing.asp

BLUE BONDS/GREEN BONDS

Bonds take many forms and have evolved over the years to align with stakeholders' needs. Green and blue bonds have been trending recently, owing to increased awareness of sustainability and ESG considerations, with many individual investors, companies and even governments stepping up their preference for such bonds. Stakeholders are increasingly concerned about the environmental impact of business operations. Carbon emissions are the main cause of climate change and global warming, which affect the agricultural business sector and the health of humans and animals. Governments, renewable-energy companies and environmental activists are keen to allocate investment, time and knowledge towards promoting a greener and sustainable economy. The United Nations Framework Convention on Climate Change (UNFCCC) 196 nations were party to the 12 December 2015 Paris Agreement, which aims to reduce greenhouse gas emissions to levels consistent with holding the increase in global average temperature to well below 2° Celsius above pre-industrial levels and pursuing efforts to limit the temperature increase to 1.5° Celsius above pre-industrial levels. Meeting these targets would require an unprecedented allocation of capital, measured in trillions of dollars a year.

Blue bonds are a type of financial instrument designed to support sustainable marine and ocean-related projects. They are similar to green bonds, which focus on environmental projects, but with a specific emphasis on ocean conservation and sustainable use of marine resources.

Blue bonds can be used to finance a variety of projects, such as:

1. Marine protected areas: Establishing and managing protected areas to conserve biodiversity and ecosystem services.
2. Sustainable fisheries: Implementing sustainable fishing practices, reducing bycatch, and promoting eco-labeling.
3. Coastal resilience: Enhancing the resilience of coastal communities to climate change, sea-level rise, and extreme weather events.
4. Ocean pollution reduction: Reducing marine pollution from plastics, agricultural runoff, and other sources.
5. Sustainable aquaculture: Promoting sustainable aquaculture practices, reducing environmental impacts, and improving social responsibility.

The benefits of blue bonds include:
1. Raising capital for ocean conservation and sustainable development
2. Promoting sustainable use of marine resources
3. Supporting climate change mitigation and adaptation efforts
4. Enhancing biodiversity and ecosystem services
5. Providing a new asset class for investors seeking sustainable investments

Examples of blue bonds include:

1. The World Bank's blue bond for the Seychelles, which raised $15 million to support marine conservation and sustainable fisheries.
2. The European Investment Bank's (EIB) blue bond for the Mediterranean, which supports sustainable fisheries and marine conservation projects.

Overall, blue bonds have the potential to play a significant role in supporting ocean conservation and sustainable development, while also providing a new investment opportunity for those seeking sustainable returns.

Source: https://img.jagranjosh.com/images/2021/January/2912021/green-bonds-india.jpg https://data.cbonds.info//files/glossary/202312/mceu_85905834011702833824160.jpg
Reference link: https://www.acuitykp.com/market-guide/financial-institutions-portfolios-and-services/ & META AI https://images.squarespace-cdn.com/content/v1/5718b643e707eb46ff2ab-c3c/1691046058485-8HCASP4UL0RAPQY0UCJ1/2.png?format=1000w

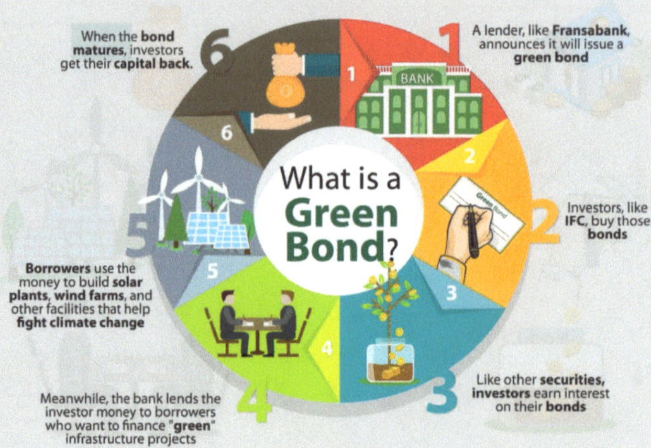

Green bonds are a type of fixed-income instrument specifically designed to finance environmentally friendly projects. They are similar to traditional bonds but with a focus on supporting sustainable development and reducing environmental impact.

Characteristics of green bonds:

1. Use of proceeds: Green bonds are issued to finance specific environmentally friendly projects, such as renewable energy, sustainable infrastructure, or green buildings.

2. Environmental impact: The projects financed by green bonds must have a positive environmental impact, such as reducing greenhouse gas emissions or promoting sustainable resource use.

3. Transparency: Issuers of green bonds must provide clear and transparent reporting on the use of proceeds and the environmental impact of the projects financed.

4. Verification: Green bonds are often verified by independent third-party organizations to ensure compliance with green bond principles.

Types of green bonds:

1. Climate bonds: Focus on projects that reduce greenhouse gas emissions or support climate change adaptation and resilience.

2. Sustainability bonds: Finance projects that promote sustainable development, including environmental, social, and governance (ESG) considerations.

3. Blue bonds: Specifically focus on marine and ocean-related projects, such as sustainable fisheries, marine conservation, and coastal resilience.

Benefits of green bonds:

1. Raising capital for sustainable projects: Green bonds provide a dedicated funding source for environmentally friendly projects.

2. Promoting sustainable development: Green bonds support projects that reduce environmental impact and promote sustainable development.

3. Enhancing transparency and accountability: Green bonds require issuers to provide transparent reporting on the use of proceeds and environmental impact.

4. Attracting socially responsible investors: Green bonds appeal to investors seeking environmentally responsible investment opportunities.

Examples of green bonds:

1. The World Bank's green bond: Issued in 2008 to finance climate-friendly projects, such as renewable energy and energy efficiency.

2. The European Investment Bank's (EIB) climate awareness bond: Issued in 2010 to finance climate-related projects, such as renewable energy and sustainable transportation.

3. The City of New York's green bond: Issued in 2019 to finance sustainable infrastructure projects, such as green roofs and energy-efficient buildings.

Overall, green bonds have become an increasingly important tool for financing sustainable development and reducing environmental impact.

GREEN FINANCE

Simply put, green finance is a loan or investment that promotes environmentally-positive activities, such as the purchase of ecologically-friendly goods and services or the construction of green infrastructure. As the hazards connected to ecologically destructive products and services rise, green finance is becoming a mainstream phenomenon.

WHY GREEN FINANCING?

Green finance delivers economic and environmental advantages to everybody. It broadens access to environmentally-friendly goods and services for individuals and enterprises, equalizing the transition to a low-carbon society, resulting in more socially inclusive growth. This results in a 'great green multiplier' effect in which both the economy and the environment gain, making it a win-win situation for everyone.

TYPES OF GREEN FINANCING

1. Green Mortgages
2. Green Loans
3. Green Credit Cards
4. Green Banks
5. Green Bonds

Source:https://bsmedia.business-standard.com/_media/bs/img/article/2023-02/28/full/1677602143-0474.jpg?im=FeatureCrop,size=(826,465)
Reference link: https//emeritus.org/blog/finance-what-is-green-finance/#what-is-green-finance

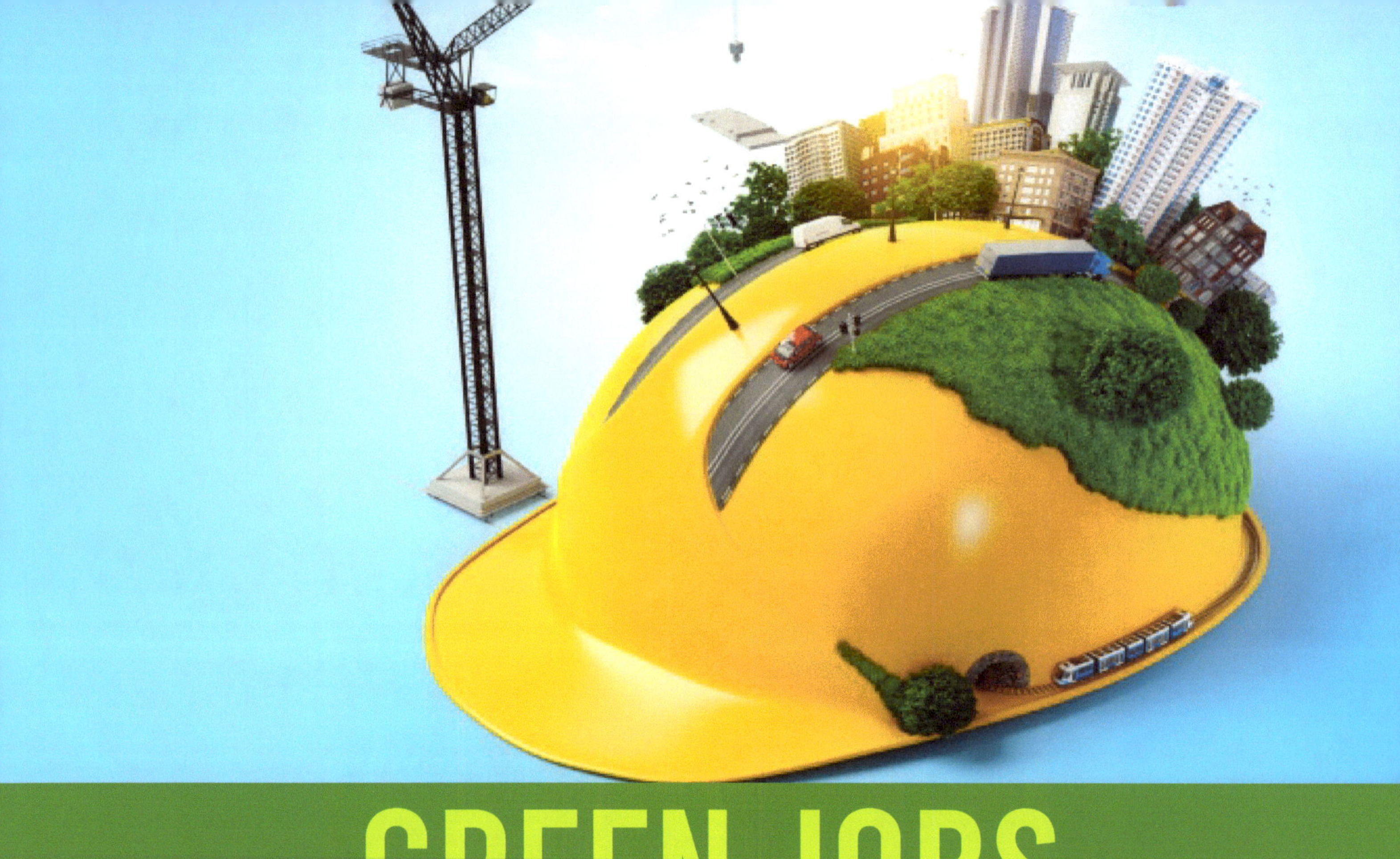

GREEN JOBS

Green jobs preserve/restore nature through renewable energy, waste management, sustainable agriculture, and green construction.

1 The Green Jobs are decent jobs that contribute to preserve or restore the environment, be they in traditional sectors such as manufacturing and construction, or production processes in environment friendly or use fewer natural reSources.

2 The Green Jobs industry includes agriculture, energy, environmental management, environmental scientist, ecologist, environmentalist, naturalist, renewable energy, conservation scientist, recycling, transportation, civil engineers, landscape architects, waste management, biologist, and many more.

Green jobs contribute to the orderly use of reSources by promoting recycling, waste reduction, and energy efficiency. These practices not only protect the environment but also save businesses and consumers money in the long run. This industry promotes skills essential for addressing environmental challenges, such as climate change, reSource conservation, pollution reduction, and sustainable reSource management

3

POLICIES AND INITIATIVES BY INDIAN GOVERNMENT

- Skill Council for Green Jobs (SCGJ)
- The National Action Plan on Climate Change (NAPCC)
- The FAME India Scheme
- Swachh Bharat Mission
- Incentives and Subsidies

Types of green jobs:
1. Renewable Energy Technicians: Install, maintain, and repair solar panels, wind turbines, and other renewable energy systems.
2. Sustainability Consultants: Help organizations develop and implement sustainable practices, reducing their environmental impact.
3. Green Building Architects: Design and develop energy-efficient, environmentally friendly buildings.
4. Environmental Scientists: Conduct research, monitoring, and analysis to understand and mitigate the impact of human activities on the environment.
5. Eco-Friendly Product Designers: Create products that are environmentally friendly, sustainable, and minimize waste.
6. Urban Farmers: Grow and distribute fresh produce in urban areas, promoting sustainable agriculture and reducing carbon footprint.
7. Climate Change Analysts: Analyze and interpret data related to climate change, helping organizations and governments develop strategies to mitigate its impacts.
8. Green Infrastructure Specialists: Design, build, and maintain green infrastructure, such as green roofs, rain gardens, and green walls.
9. Sustainable Transportation Specialists: Develop and implement sustainable transportation systems, promoting alternative modes of transportation and reducing emissions.
10. Waste Management Specialists: Develop and implement sustainable waste management practices, reducing waste and promoting recycling.

Source: https://trellis.net/wp-content/uploads/2024/07/greenjobs_vadim_georgiev_sstock.jpg Reference link: https://earth5r.org/the-evolving-green-job-sector-of-india/ & Meta AI

GREEN CLIMATE FUND

The Green Climate Fund (GCF) is a new global fund created by the United Nations Framework Convention on Climate Change to support the efforts of developing countries to respond to the challenge of climate change. The Green Climate Fund (GCF) is the world's largest environmental fund that seeks to help developing nations in cutting down their greenhouse gas emissions, while at the same time making them adapt suitably to climate change.

This is done by supporting projects, programmes, policies and other activities through a state-of-the-art funding window.It was established formally as a financing mechanism by the United Nations Framework Convention on Climate Change (UNFCCC) in 2010 it is headquartered in the Songdo district in South Korea.

The mandate of GCF is to promote a paradigm shift towards low-emission and climate-resilient development pathways by providing support to developing countries to limit or reduce their greenhouse gas emissions (mitigation) and to adapt to the impacts of climate change (adaptation). Given the urgency and seriousness of climate change, the purpose of GCF is to make a significant and ambitious contribution to global efforts towards attaining the goals set by the international community to combat climate change.

Source:https://www.rural21.com/fileadmin/_processed_/5/f/csm_News_4_The_Green_Climate_Fund-photo_f1faa35f96.jpg Reference link: https://byjus.com/free-ias-prep/green-climate-fund/

GREEN HYDROGEN

Green hydrogen is the hydrogen produced by splitting water into hydrogen and oxygen using renewable electricity through a process called electrolysis. This results in very low or zero carbon emissions. Emerging green hydrogen strategies and policies differ widely on the definition of "renewable energy", the boundaries of the carbon accounting system, the emission thresholds at which hydrogen is considered green, and the feedstocks and production technologies deployed. This lack of standardisation is undermining efforts to accelerate the use of green hydrogen.

NEED FOR PRODUCING GREEN HYDROGEN:

Hydrogen is a great Source of energy because of its high energy content per unit of weight, which is why it is used as rocket fuel.

Green hydrogen in particular is one of the cleanest Sources of energy with close to zero emission. It can be used in fuel cells for cars or in energy-guzzling industries like fertilizers and steel manufacturing.

Countries across the world are working on building green hydrogen capacity as it can ensure energy security and also help in cutting carbon emission.

Green hydrogen has become a global buzzword, especially as the world is facing its biggest-ever energy crisis and the threat of climate change is turning into a reality.

Source:https://s7d1.scene7.com/is/image/wbcollab/hero_two_hydro?qlt=90&fmt=webp&resMode=sharp2
Reference link: https://www.drishtiias.com/daily-updates/daily-news-analysis/national-green-hydrogen-mission-1

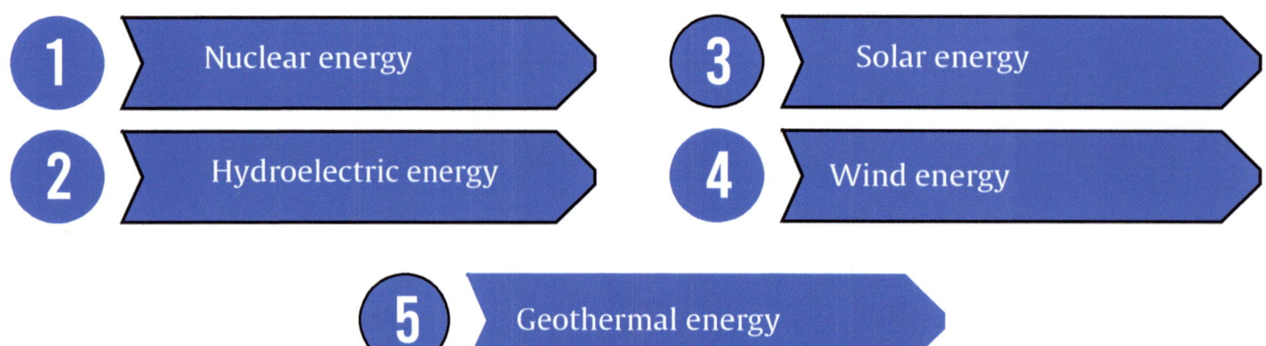

CLEAN ENERGY

The term "clean" or "carbon-free" energy is used to refer to the electricity that is generated by facilities that do not directly emit greenhouse gases such as carbon dioxide during the generating process. Though there is some overlap between the categories, clean energy is different from "green" energy and "renewable" energy. Clean energy production allows us to generate the energy we need without the greenhouse gas emissions and negative environmental effects that come with fossil fuels, in turn helping to reduce climate change.

EXAMPLES OF CLEAN ENERGY

1. Nuclear energy

2. Hydroelectric energy

3. Solar energy

4. Wind energy

5. Geothermal energy

Source: https://etimg.etb2bimg.com/photo/101156859.cms Reference link: https://www.constellation.com/energy-101/energy-innovation/what-is-clean-energy.html

CLEAN AIR

The National Clean Air Programme is a pollution-control initiative that was started by the Ministry of Environment, Forest and Climate Change (MoEFCC) in 2019 with the goal of reducing the concentration of coarse (particulate matter with a diameter of less than 10 micrometres, or PM10) and fine (particulate matter with a diameter of less than 2.5 micrometres, or PM2.5) particles by at least 20% over the course of the following five years, using 2017 as the baseline year for comparison. The National Clean Air Programme (NCAP) of the Government of India is a significant step in recognizing and resolving the issue of declining ambient air quality. With an emphasis on about 132 "non-attainment" cities whose air quality requirements are not being reached, the NCAP has set a deadline for improving air quality across the nation. The NCAP offers cities a comprehensive framework for creating air quality management plans as well as recommendations for policies in several fields. Based on the recommendations of the 15th Finance Commission, the Government of India allocated roughly \$1.7 billion in 2020 to combat air pollution for the 42 Indian cities with a million or more inhabitants over the following five years – provided they reduce their air pollution levels by 15% annually. This is the first fiscal transfer funding program for managing urban air quality that is based on performance. In August 2021, the Indian Parliament enacted a statute to create the Commission of Air Quality Management in the National Capital Region and surrounding districts. This was done in recognition of the necessity for coordinated cross-jurisdictional and airshed level action and coordination.

Source: https://st.depositphotos.com/1252160/4918/i/450/depositphotos_49183437-stock-photo-engineer-try-to-make-a.jpg Reference link: https://www.impriindia.com/insights/policy-update/national-clear-air-programme/

CLEAN TRANSPORTATION

Green transportation is any means of travel that doesn't negatively impact the environment. Green transportation can be private (like a fast ebike), or public transit (like an electric city bus). Walking is also considered a green mode of travel.

The common denominator among green transportation everywhere is that it's sustainable. Sustainable transportation is powered by reSources that aren't depleted when used, making them harnessable by future generations.

An efficient Transport Sector is important for economic development of the country and for the well-being of its people. The transport sector makes up 30% of the global energy consumption. Its energy use is expected to grow 1% every year till 2030.

In India, the transport sector has grown extensively, both in terms of physical spread as well as capacity to meet the mobility demands for both passengers as well as freight. Despite its impressive growth, it is seen that the existing transport infrastructure in India is far from meeting the growing mobility needs in terms of coverage, capacity as well as service quality.

Unsustainable transport activities can produce widespread negative impacts like degradation of air quality, greenhouse gas emissions, increased threat of global climate change and habitat loss of animals and fragmentation. Therefore, there is a need to pay greater attention to sustainable (green) transport at city, state and national level as the way forward for India's mobility sector.

Source: https://www.shutterstock.com/image-vector/electric-transportation-green-alternative-energy-600nw-1878519067.jpg Reference link: https://keegomobility.com/blog/what-is-green-transportation/ https://www.drishtiias.com/daily-updates/daily-news-editorials/india-s-transition-to-green-transport

CLEAN ELECTRICITY TAX

The clean electricity tax typically refers to a financial incentive or tax policy aimed at promoting the use of renewable energy Sources and reducing reliance on fossil fuels. It can take various forms, such as:

TAX CREDITS	These are often provided to individuals or businesses that invest in renewable energy technologies, like solar panels or wind turbines.
TAX DEDUCTIONS	Tax deductions may be available for expenses related to the installation or maintenance of clean energy systems.
CARBON TAX	This is a tax on carbon emissions, which encourages companies to adopt cleaner energy Sources to avoid higher costs.
SUBSIDIES AND GRANTS	While not a tax per se, governments may offer financial assistance to encourage clean energy investments.

THE CLEAN DEVELOPMENT MECHANISM

The Clean Development Mechanism (CDM), defined in Article 12 of the Protocol, allows a country with an emission-reduction or emission-limitation commitment under the Kyoto Protocol (Annex B Party) to implement an emission-reduction project in developing countries. Such projects can earn saleable certified emission reduction (CER) credits, each equivalent to one tonne of CO_2, which can be counted towards meeting Kyoto targets.

The mechanism is seen by many as a trailblazer. It is the first global, environmental investment and credit scheme of its kind, providing standardized emissions offset instrument, CERs. A CDM project activity might involve, for example, a rural electrification project using solar panels or the installation of more energy-efficient boilers. The mechanism stimulates sustainable development and emission reductions, while giving industrialized countries some flexibility in how they meet their emission reduction or limitation targets.

Source: https://frostbrowntodd.com/app/uploads/2023/03/clean-renewable-energy-or-electricity-production-tax-credits-and-incentives-financial-concept-green-energy-symbols-atop-coin-stack-eg-solar-panel-wind-turbine-fuel-cell-battery-and-the-word-tax-stockpack-gettyimages-scaled.jpg Reference link: ChatGPT

qMK2K-HBJlap1a24 Reference link: https://unfccc.int/process-and-meetings/the-kyoto-protocol/mechanisms-under-the-kyoto-protocol/the-clean-development-mechanism

Source: https://media.licdn.com/dms/image/D4E12AOG9O2SpBeNYuKg/article-cover_image-shrink_720_1280/0/1655640712803?e=2147483647&v=beta&t=DZKDH2OKILRttnCLBXIlBUCY5-cx-

CARBON FOOTPRINT

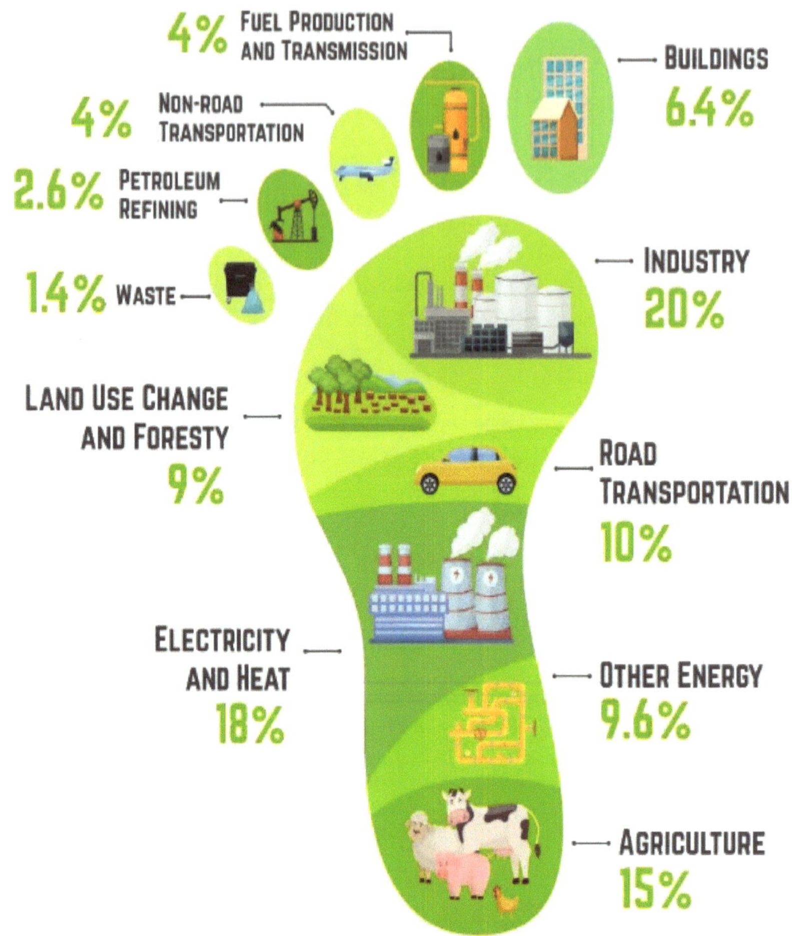

CARBON FOOTPRINT (SCOPE – 1, SCOPE – 2, SCOPE - 3)

A carbon footprint is the total amount of greenhouse gases (including carbon dioxide and methane) that are generated by our actions. The average carbon footprint for a person in the United States is 16 tons, one of the highest rates in the world. Globally, the average carbon footprint is closer to 4 tons. To have the best chance of avoiding a 2 rise in global temperatures, the average global carbon footprint per year needs to drop to under 2 tons by 2050.

Lowering individual carbon footprints from 16 tons to 2 tons doesn't happen overnight! By making small changes to our actions, like eating less meat, taking fewer connecting flights and line drying our clothes, we can start making a big difference. According to the U.S. EPA, transportation and electricity—two essential functions for most businesses—produce over half of greenhouse gas emissions (GHG emissions) in the U.S. Both are great places to start if you're looking to mitigate your GHG emissions, but did you know that global emissions are about more than what comes out of an exhaust pipe or a smokestack?

If you honestly care about sustainability, you must measure and track your carbon emissions before attempting to reduce your carbon footprint. But you have to look across your entire busi ness, and emissions scopes—often colloquially referred to as "scope emissions" or scope 1, 2, and 3—help you break down your emissions Sources and behaviors. While CO_2 emissions scopes might seem confusing at first, they actually help you create a GHG inventory by identifying your total emissions or how much CO_2e you emit based on everything it takes for your business to operate. What do the different emissions scopes mean?

THE GREENHOUSE GAS PROTOCOL (GHG PROTOCOL) DIVIDES EMISSIONS INTO THREE SCOPES

SCOPE 1 EMISSIONS – direct emissions from Sources owned or controlled by a company

SCOPE 2 EMISSIONS – indirect emissions from purchased electricity, steam, heat, and cooling

SCOPE 3 EMISSIONS – all other emissions associated with a company's activitiesIf this is hard to grasp at first,

we have a good shorthand to remember what each scope includes:
Burn, Buy, Beyond. Scope 1 is what you burn; scope 2 is energy you buy; and scope 3 is everything beyond that.While scope 1 and scope 2 emissions might be the easiest to measure, tracking what is often the largest culprit of a company's carbon footprint—scope 3 emissions—tend to be more nebulous. Scope 3 emissions include an array of elusive carbon dioxide-emitting activities that, when added up, often account for more significant carbon emissions than Scopes 1 and 2 combined. If a company truly intends to reduce or even eliminate its carbon footprint, it must address all three scopes and pay special attention to scope 3.

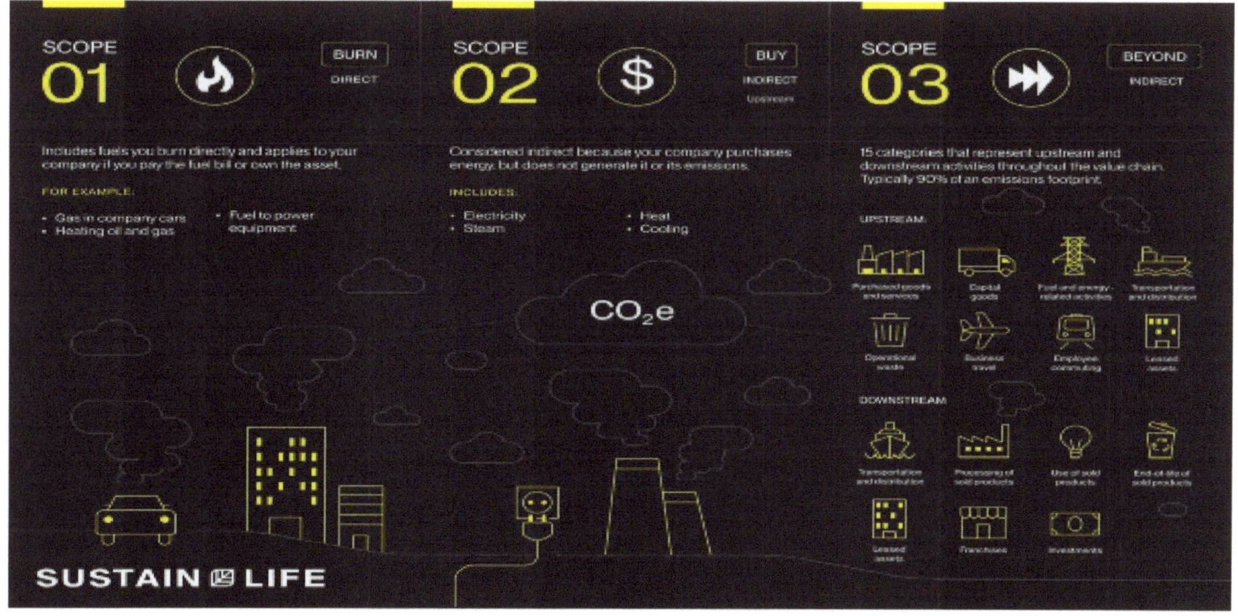

Source: https://www.daikin.co.uk/adobe/dynamicmedia/deliver/dm-aid--048044b2-897e-4885-a506-e9da6982f74c/how-can-i-reduce-my-carbon-footprint.jpg?preferwebp=true&quality=90 Reference link: https://www.nature.org/en-us/get-involved/how-to-help/carbon-footprint-calculator/ https://www.sustain.life/blog/scope-emissions

CARBON-INTENSITY

Carbon intensity is a measure of how clean our electricity is. It refers to how many grams of carbon dioxide (CO_2) are released to produce a kilowatt hour (kWh) of electricity.

Electricity that's generated using fossil fuels is more carbon intensive, as the process by which it's generated creates CO_2 emissions.

Renewable energy Sources, such as wind, hydro or solar power, produce next to no CO_2 emissions, so their carbon intensity value is much lower and often zero.

Using electricity with a low carbon intensity value will reduce carbon emissions overall – especially if we use it during times when the largest amounts of clean electricity are being generated.

Source: https://www.advancedsciencenews.com/wp-content/uploads/2020/08/shutterstock_130778297.jpg Reference link: https://www.nationalgrid.com/stories/energy-explained/what-is-carbon-intensity

How do Carbon Footprints Affect the Economy

CARBON INTENSITY OF THE ECONOMY

Emission intensity (also carbon intensity or C.I.) is the emission rate of a given pollutant relative to the intensity of a specific activity, or an industrial production process; for example grams of carbon dioxide released per mega joule of energy produced, or the ratio of greenhouse gas emissions produced to gross domestic product (GDP). Emission intensities are used to derive estimates of air pollutant or greenhouse gas emissions based on the amount of fuel combusted, the number of animals in animal husbandry, on industrial production levels, distances traveled or similar activity data. Emission intensities may also be used to compare the environmental impact of different fuels or activities. In some case the related terms emission factor and carbon intensity are used interchangeably. The jargon used can be different, for different fields/industrial sectors; normally the term "carbon" excludes other pollutants, such as particulate emissions. One commonly used figure is carbon intensity per kilowatt-hour (CIPK), which is used to compare emissions from different Sources of electrical power.

The carbon Intensity of electricity measures the amount of greenhouse gases emitted per unit of electricity produced. The units are in grams of CO_2 equivalents per kilowatt-hour of electricity. Carbon emission intensity of economies in kg of CO_2 per unit of GDP (2016)

CARBON OFFSETS/OFFSETTING

Carbon offsets are tradable "rights" or certificates linked to activities that lower the amount of carbon dioxide (CO_2) in the atmosphere. By buying these certificates, a person or group can fund projects that fight climate change, instead of taking actions to lower their own carbon emissions. In this way, the certificates "offset" the buyer's CO_2 emissions with an equal amount of CO_2 reductions somewhere else.

HOW DOES BUYING CARBON OFFSETS KEEP CO_2 OUT OF THE ATMOSPHERE?

Carbon offsets fund specific projects that either lower CO_2 emissions, or "sequester" CO_2, meaning they take some CO_2 out of the atmosphere and store it. Some common examples of projects include reforestation, building renewable energy, carbon-storing agricultural practices, and waste and landfill management. Reforestation in particular is one of the most popular types of projects to produce carbon offsets. Carbon offsets are granted to project owners, who sell them to third parties like companies that want to balance the CO_2 they put into the atmosphere by paying to remove CO_2 from somewhere else. Although carbon offsets are easy to understand, there are many challenges in producing them. To issue carbon offsets, a project needs to prove it will actually reduce emissions. The amount of CO_2 being kept out of the atmosphere also needs to be accurately measured.

This process requires well-documented standards and protocols, as well as a trusted way to verify that the project is doing everything it claims. These procedures can be expensive and specific to one type of project. But without them, we cannot trust that buying carbon offsets really lowers the amount of CO_2 in the atmosphere.

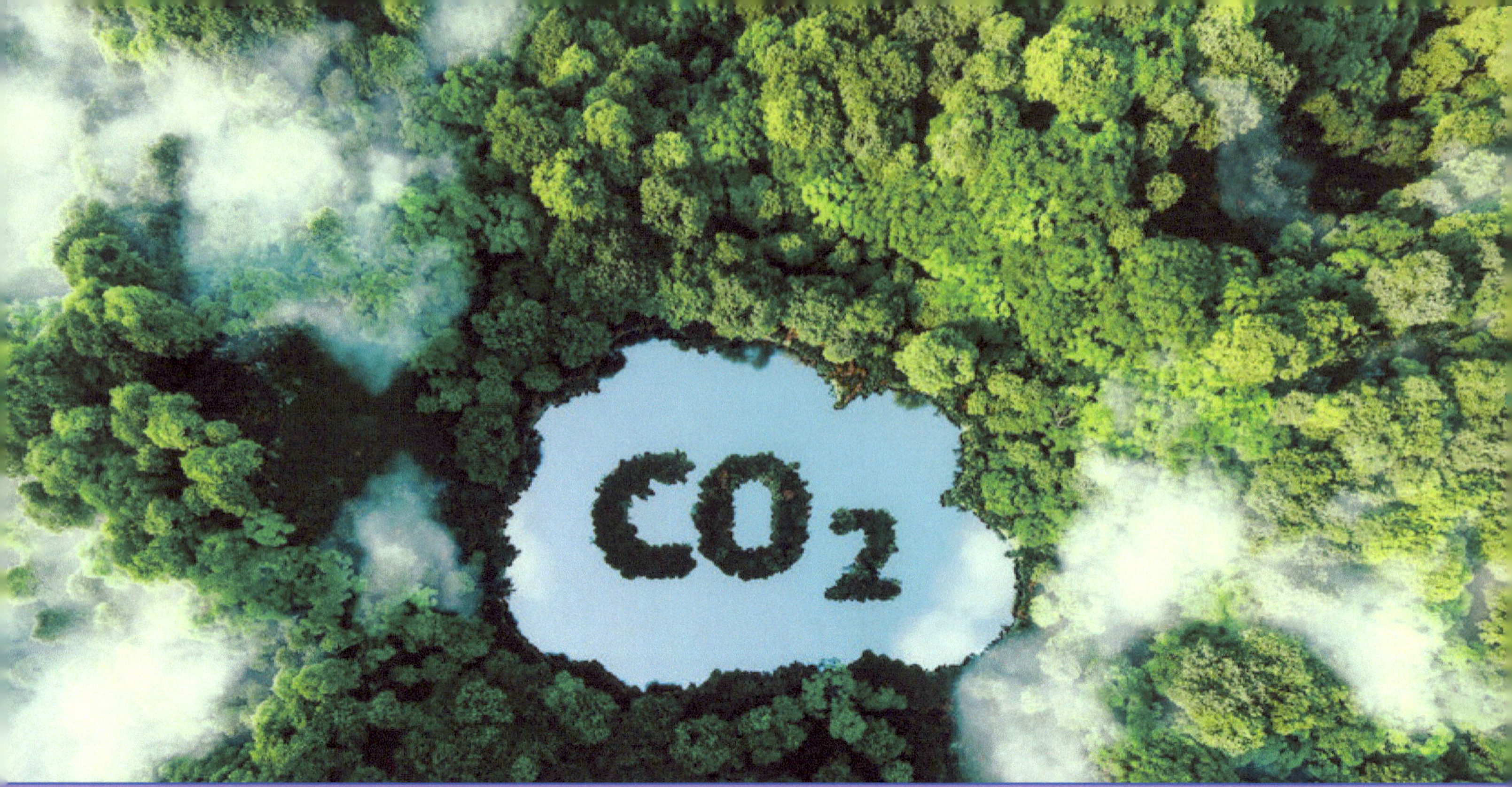

CARBON OFFSET CREDITS

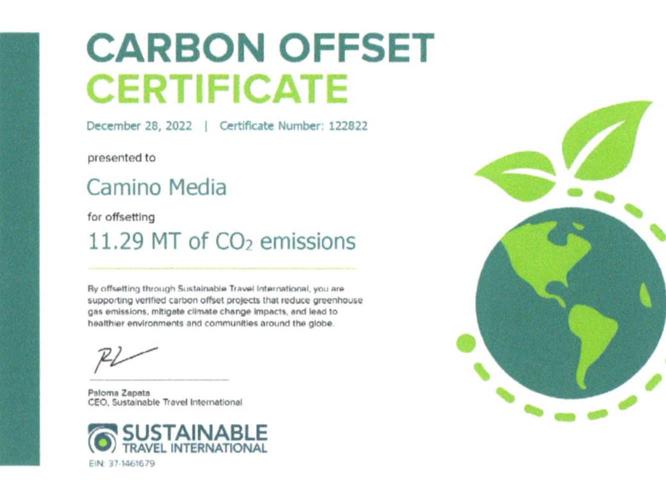

CARBON OFFSET
CERTIFICATE

December 28, 2022 | Certificate Number: 122822

presented to

Camino Media

for offsetting

11.29 MT of CO_2 emissions

By offsetting through Sustainable Travel International, you are supporting verified carbon offset projects that reduce greenhouse gas emissions, mitigate climate change impacts, and lead to healthier environments and communities around the globe.

Paloma Zapata
CEO, Sustainable Travel International

SUSTAINABLE
TRAVEL INTERNATIONAL
EIN: 37-1461679

Carbon offset credits are permits that represent a reduction in greenhouse gas emissions, typically measured in metric tons of CO_2 equivalent. When a company or individual purchases these credits, they're essentially funding projects that either reduce emissions (like renewable energy projects, reforestation, or energy efficiency initiatives) or remove carbon from the atmosphere (such as carbon capture and storage).

The idea is that by offsetting your carbon footprint—whether from travel, energy use, or other activities you can contribute to the fight against climate change. One credit usually offsets one ton of CO_2, and these credits are often traded in markets, allowing companies to meet regulatory requirements or voluntarily reduce their impact on the environment.

Source: https://www.corporateknights.com/wp-content/uploads/2023/02/174384169_l.jpg Reference link: ChatGPT , https://www.productionservicenetwork.com/wp-content/uploads/STI_CarbonOffsetCertificate_2022.jpg

CARBON CREDITS

Carbon credit is a tradable instrument (typically a virtual certificate) that conveys a claim to avoided GHG emissions or to the enhanced removal of GHG from the atmosphere. Credits allow claims to be transferred from an entity that generated the avoided emissions or enhanced removals to a buyer. The buyer of a carbon credit can then "retire" it to count the avoided emissions or enhanced removals towards a climate change mitigation goal.

Carbon credits are certified by either governments or independent certification bodies (also known as "carbon crediting programs"). A single credit is typically denominated to represent the equivalent of one metric tonne of CO_2 avoided or removed (see Box 1). The terms carbon offset credits, carbon offsets, offset credits or simply offsets may be used interchangeably, though carbon credits are the preferred technical term as the credit is what is used for compliance or voluntary reporting purposes, and carbon credits, as opposed to offset credits, are not readily confused with the verb "to offset" and the practice of offsetting.

CARBON SINK

In the fight against climate change, not only humanstry to counteract the effects of global warming with mitigation and adaptation measures, but nature itself has its own weapons to try to keep the average temperature of the planet from increasing. For that, carbon sinks, which are natural (oceans and forests) and artificial deposits (certain technologies and chemicals) absorb and capture carbon dioxide (CO_2) from the atmosphere and reduce its concentration in the air.

NATURAL CARBON SINKS: OCEANS AND FORESTS

Oceans are considered the main natural carbon sinks, as they are capable of absorbing about 50% of the carbon emitted into the atmosphere. In particular, plankton, corals, fish, algae and other photosynthetic bacteria are responsible for this capture. Oceans are the main carbon sinks and absorb up to 50% of CO_2 In the case of forests and other woodland areas, carbon sequestration is done through photosynthesis. Plants absorb CO_2 from the atmosphere, store some of its carbon content of, and return oxygen to the atmosphere. The problem with natural carbon sinks is that they have a maximum limit, causing (among other impacts) ocean acidification when exceeded. This acidification consists of a decrease of the pH of the water caused by the absorption of carbon dioxide. The acidification of the oceans negatively impacts species such as corals, algae, shellfish and mollusks, which are weakened and in many cases become ill and die.

ARTIFICIAL TECHNIQUES FOR CARBON SEQUESTRATION

To enhance and accelerate the natural process of carbon sequestration there are artificial techniques that extract carbon from the atmosphere and store it in the earth's crust. However, these technologies have not acquired the efficiency and maturity needed to cope with the extreme changes that climate change poses, and sometimes, in critical cases, CO_2 escapes the artificial sinks (carbon leakage)

In conclusion, carbon sinks are an important help to combat climate change, but they do not solve it. It is essential to abandon our dependence on fossil fuels and to make a strong commitment to renewable energies.

Source: https://cdn.sketchbubble.com/pub/media/catalog/product/optimized1/2/b/2b4a6f005c318deb4395eb8856bdde959fdc6533a1e6cf32799ba1d65bd400bb/carbon-sinks-slide1.png
Reference link: https://www.activesustainability.com/climate-change/carbon-sinks-what-are/?_adin=11734293023

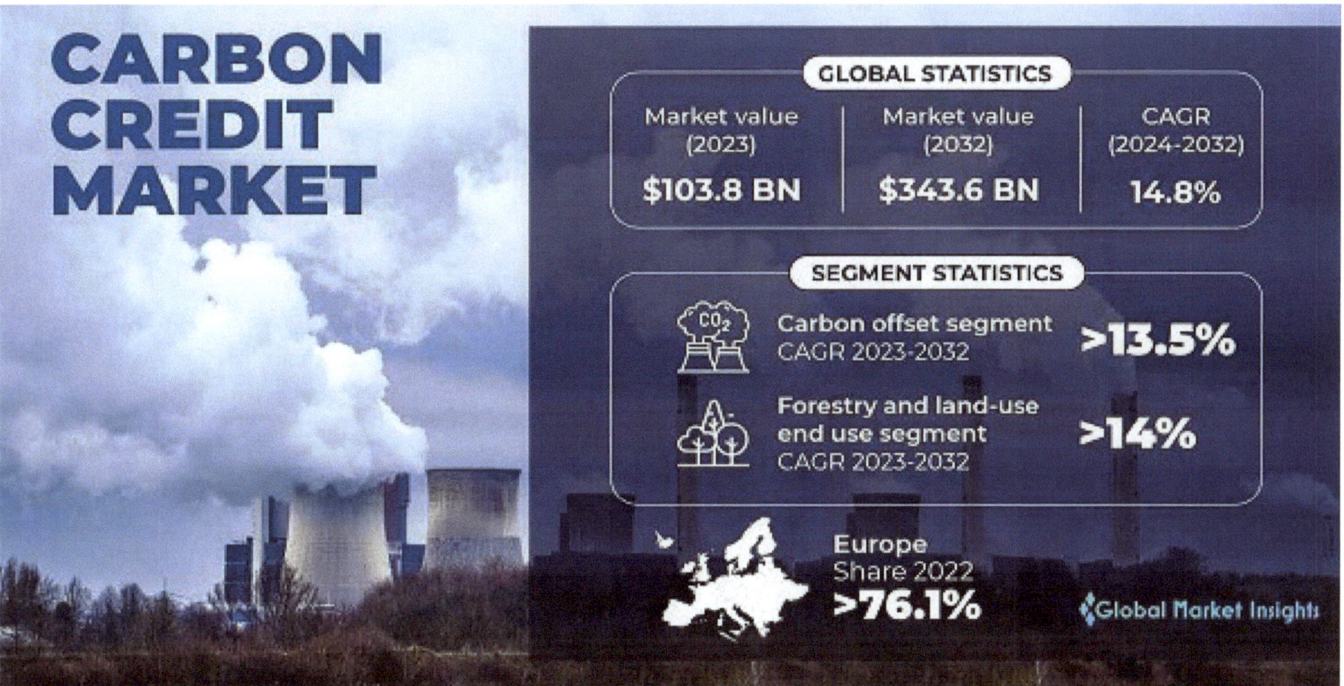

The carbon credit market is a key mechanism aimed at reducing greenhouse gas emissions. A carbon credit represents one metric ton of carbon dioxide (CO_2) or an equivalent amount of other greenhouse gases that is either reduced, avoided, or removed from the atmosphere.

TYPES

THERE ARE TWO MAIN TYPES OF CARBON CREDITS:

1 Compliance Credits — These are used in regulatory frameworks (like cap-and-trade systems) where companies must meet legally binding emissions limits.

2 Voluntary Credits — These are bought by companies or individuals who want to offset their emissions voluntarily.

HOW THE MARKET WORKS

1 Cap-and-Trade Systems — Governments set a cap on total emissions and distribute or auction off carbon credits. Companies that reduce emissions can sell their excess credits to others that exceed their limits.

2 Voluntary Markets — Companies and individuals can purchase carbon credits from projects that reduce or remove emissions, such as reforestation or renewable energy projects, even if they are not legally required to do so.

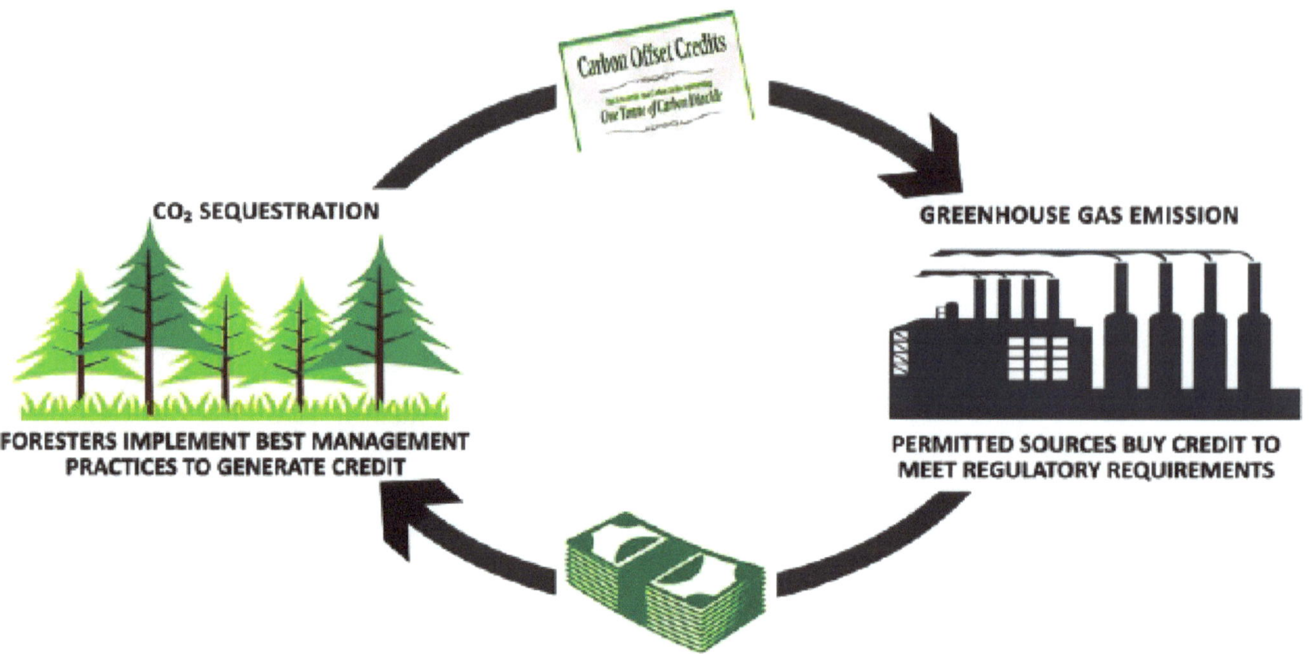

CARBON CREDIT MECHANISM (PARIS AGREEMENT CREDITING MECHANISM PACM)

Article 6 of the Paris Agreement sets out how countries can pursue voluntary cooperation to reach their climate targets. It enables international cooperation to tackle climate change and unlock financial support for developing countries. This means that, under Article 6, countries are able to transfer carbon credits earned from the reduction of greenhouse gas emissions to help one or more countries meet their climate targets. There are three tools which countries can draw upon under Article 6, one of which is the Paris Agreement Crediting Mechanism (PACM) - the UN's new high-integrity carbon crediting mechanism.

HOW DOES THE PARIS AGREEMENT CREDITING MECHANISM WORK?

The carbon crediting mechanism under the Paris Agreement allows countries to raise climate ambition and implement national action plans more affordably.It identifies and encourages opportunities for verifiable emission reductions, attracts funding to implement them, and allows cooperation among countries and other groups to conduct and benefit from these activities.

For example, through this mechanism a company in one country can reduce emissions in that country and have those reductions credited, so that it can sell them to another company in another country. That second company may use them for complying with its own emission reduction obligations or to help it meet net-zero targets.The Paris Agreement Crediting Mechanism can also be a Source of climate finance for developing nations, with a share of proceeds going towards adaptation funding to build resilience to the inevitable impacts of climate change.

Source: https://pwonlyias.com/wp-content/uploads/2024/07/unnamed-2024-07-23t121704470-669f9cce34a29.webp
Reference link: https://unfccc.int/process-and-meetings/the-paris-agreement/article-64-mechanism

CARBON REDUCTION

arbon reduction refers to the decrease or minimization of carbon dioxide (CO_2) emissions, which contribute to global warming and climate change. It involves reducing the amount of greenhouse gases released into the atmosphere, primarily through decreasing fossil fuel consumption, increasing energy efficiency, transitioning to renewable energy Sources, implementing carbon capture and storage technologies, promoting sustainable land use and forestry practices. The goal of carbon reduction is to mitigate climate change by lowering the amount of CO_2 and other greenhouse gases in the atmosphere, thereby reducing the Earth's temperature increase.

HOW TO REDUCE YOUR CARBON FOOTPRINT THROUGH TRANSPORTATION

1. Drive less
2. Go easy on the acceleration and brakes
3. Regularly service your car and keep tires properly inflated

HOW TO REDUCE YOUR CARBON FOOTPRINT THROUGH FOOD

1. Eat less meat and stick with fruits, veggies, grains and beans
2. Choose organic and local foods that are in season
3. Reduce your food waste

HOW TO REDUCE YOUR CARBON FOOTPRINT AT HOME

1. Turn down your water heater to 120°F
2. Lower your thermostat in winter and raise it in summer
3. Turn off lights and unplug appliances when not in use

HOW TO REDUCE YOUR CARBON FOOTPRINT WHEN SHOPPING

1. Bring a reusable bag
2. Invest in quality products that last

Source: https://www.planetmark.com/wp-content/uploads/2021/11/7-ways-businesses-can-reduce-carbon-emissions.jpg
Reference link:ChatGPT https://www.constellation.com/energy-101/energy-innovation/how-to-reduce-your-carbon-footprint.html

CARBON CREDIT SCHEME

The United Nations allows countries a certain number of credits, and each nation is responsible for issuing, monitoring, and reporting its carbon credit status annually. Governments allow companies to emit a set amount of GHGs before needing to purchase credits. If emissions exceed limits, they are required to buy credits. If a company purchases too many credits, it can sell the excess on a carbon exchange or marketplace. This system is commonly called a cap-and-trade program.

U.S. CARBON CREDITS - Cap-and-trade programs remain controversial in the United States, but 13 states have adopted such market-based approaches to reducing greenhouse gases, according to the Center for Climate and Energy Solutions. Eleven of them are Northeast states that banded together to jointly attack the problem through a program known as the Regional Greenhouse Gas Initiative (RGGI).

CALIFORNIA'S CAP-AND-TRADE PROGRAM - The state of California initiated a cap-and-trade program in 2013. The rules apply to the state's large electric power plants, industrial plants, and fuel distributors. The state claims that its program is the fourth largest in the world after those of the European Union, South Korea, and China.

THE U.S. CLEAN AIR ACT - The United States has been regulating airborne emissions since the passage of the U.S. Clean Air Act of 1990. The act is credited as the world's first cap-and-trade program, although it calls its caps "allowances." The program is credited by the Environmental Defense Fund for substantially reducing emissions of sulfur dioxide from coal-fired power plants, the cause of the notorious acid rain of the 1980s.

THE INFLATION REDUCTION ACT - The Inflation Reduction Act is a landmark bill that was signed into law on Aug. 16, 2022. It aims to reduce the deficit, fight inflation, and reduce carbon emissions. It's hoped that these more generous credits will convince investors to make a bigger effort at capturing carbon. The previous tax incentive, known as 45Q, was accused of only paying enough to make easy carbon capture projects worth pursuing.

CARBON TRADING SCHEME

Carbon trade is the buying and selling of credits that permit a company or other entity to emit a certain amount of carbon dioxide or other greenhouse gases. The carbon credits and the carbon trade are authorized by governments with the goal of gradually reducing overall carbon emissions and mitigating their contribution to climate change.

Carbon trading scheme(Emission trading), as set out in Article 17 of the Kyoto Protocol, allows countries that have emission units to spare - emissions permitted them but not "used" - to sell this excess capacity to countries that are over their targets. Thus, a new commodity was created in the form of emission reductions or removals. Since carbon dioxide is the principal greenhouse gas, people speak simply of trading in carbon. Carbon is now tracked and traded like any other commodity. This is known as the "carbon market.

CARBON EMISSIONS

Carbon dioxide emissions are the primary driver of global climate change. It's widely recognized that to avoid the worst impacts of climate change, the world needs to urgently reduce emissions. But, how this responsibility is shared between regions, countries, and individuals has been an endless point of contention in international discussions. This debate arises from the various ways in which emissions are compared: as annual emissions by country; emissions per person; historical contributions; and whether they adjust for traded goods and services. These metrics can tell very different stories.

Source: https://images.indianexpress.com/2023/09/climate-change.jpg?w=414 Reference link: https://www.investopedia.com/terms/c/carbontrade.asp https://unfccc.int/process/the-kyoto-protocol/mechanisms/emissions-trading Source: https://akm-img-a-in.tosshub.com/indiatoday/images/story/202311/un-report-hails-indias-low-carbon-emissions-but-shows-serious-danger-ahead-245339282-16x9_0.jpg?VersionId=LYLmE7kbClcZ24ec3Rzwto9smroBWpFz Reference link: https://ourworldindata.org/co2-emissions

CARBON NEUTRAL / NEUTRALITY

According to the European Parliament, carbon neutrality is reached when the same amount of CO_2 is released into the atmosphere as is removed by various means, leaving a zero balance, also known as a zero carbon footprint. But what exactly do we mean by carbon footprint? This is defined as the total amount of GHG emissions caused by an individual, organisation, service or product. There are a number of ways of achieving the balance we are talking about. The healthiest way is not to emit more CO_2 than can be absorbed naturally by the world's forests and plants, which act as carbon sinks through the process of photosynthesis - they take in CO_2 from the air and turn it into oxygen - helping to reduce emissions. In a statement delivered in 2020, UN Secretary-General António Guterres established the key factors involved in reaching climate neutrality.

Building a true global coalition in support of carbon neutrality by 2050.

Aligning global finance behind the Paris Agreement and the Sustainable Development Goals (SDGs).

Making decisive progress on adaptation and resilience to climate change.

Source: https://terrapass.com/wp-content/uploads/2021/09/carbon-negative-illustration-comparison-with-carbon-neutral.jpg
Reference link:https://www.iberdrola.com/sustainability/what-is-carbon-neutrality & metqa AI

Carbon neutrality refers to the state of achieving net-zero carbon emissions. This means that the amount of greenhouse gas (GHG) emissions produced is equal to the amount of GHG emissions reduced or offset.

Carbon neutrality can be achieved through:

1. Reducing emissions: Implementing energy-efficient practices, switching to renewable energy sources, and reducing waste.

2. Offsetting emissions: Investing in projects that reduce GHG emissions, such as reforestation, wind farms, or carbon capture and storage.

3. Carbon removal: Using technologies that remove CO_2 from the atmosphere, such as direct air capture or afforestation/reforestation.

Benefits of carbon neutrality:

1. Mitigating climate change: Reducing GHG emissions helps slow global warming and its associated impacts.

2. Cost savings: Energy-efficient practices and renewable energy sources can reduce energy costs.

3. Enhanced brand reputation: Companies that achieve carbon neutrality demonstrate their commitment to sustainability and environmental responsibility.

4. Regulatory compliance: Carbon neutrality can help organizations comply with emerging climate regulations and policies.

Examples of carbon-neutral initiatives:

1. Countries: Norway, Sweden, and Costa Rica have set targets to become carbon neutral.

2. Companies: Patagonia, REI, and IKEA have made commitments to achieve carbon neutrality.

3. Cities: Cities like Copenhagen, Vancouver, and San Francisco have set goals to become carbon neutral.

To achieve carbon neutrality, individuals and organizations can:

1. Conduct a carbon footprint analysis: Measure and assess GHG emissions.

2. Develop a carbon reduction plan: Set targets and implement strategies to reduce emissions.

3. Invest in renewable energy: Transition to renewable energy sources like solar, wind, or hydroelectric power.

4. Offset emissions: Invest in carbon offset projects or purchase carbon credits.

5. Monitor and report progress: Regularly track and report emissions reductions and offsetting efforts.

CARBON LITERACY

Carbon Literacy is the awareness of climate change and the climate impacts of humankind's everyday actions. The term has been used in a range of contexts in scientific literature and in casual usage, but is most associated with The Carbon Literacy Project (CLP). Carbon Literacy is the knowledge and capacity required to create a positive shift in how humankind lives, works and behaves in response to climate change. The Carbon Literacy Project defines Carbon Literacy as "an awareness of the carbon costs and impacts of everyday activities and the ability and motivation to reduce emissions, on an individual, community and organizational basis."

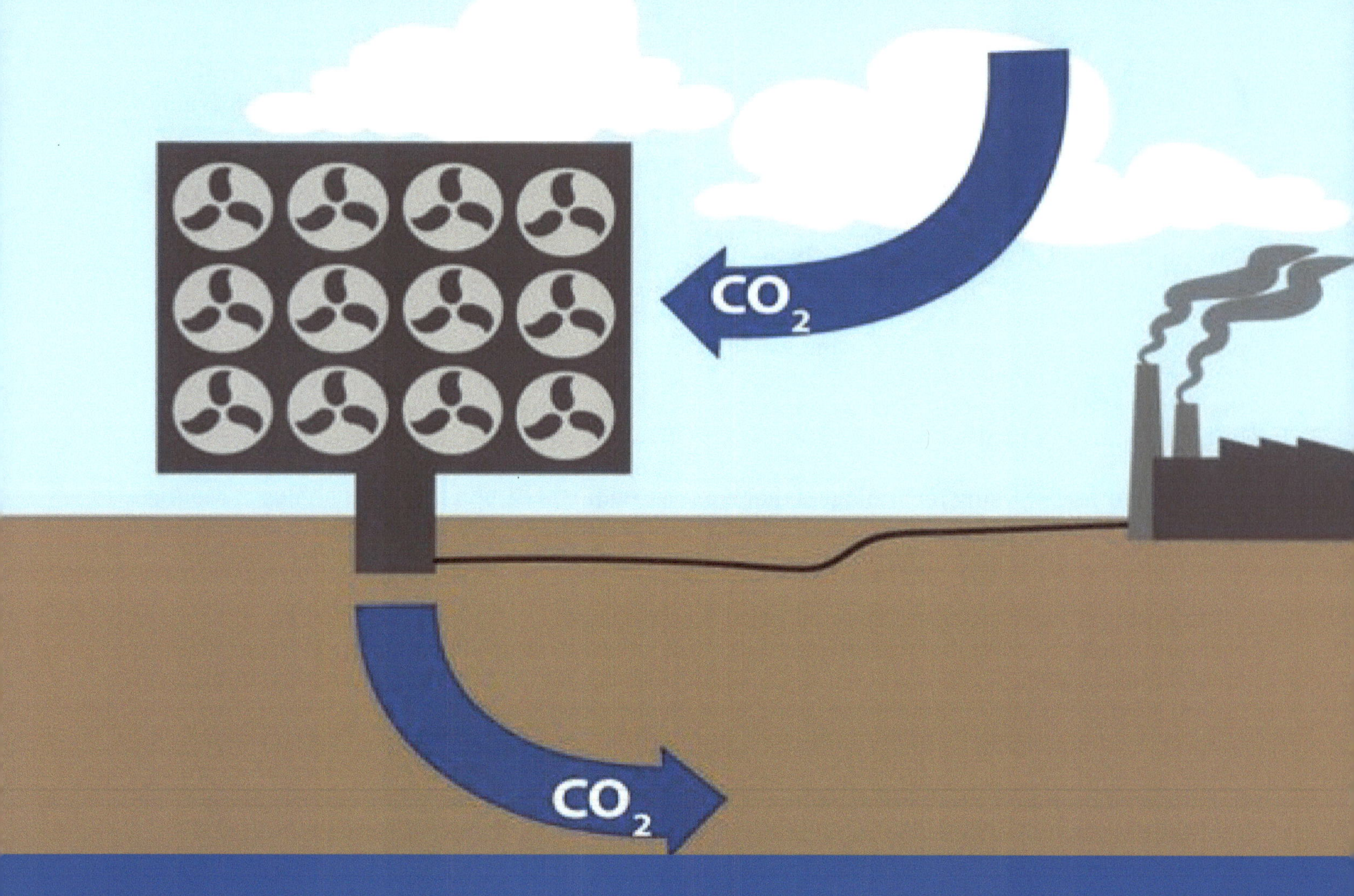

CARBON REMOVAL TECHNOLOGIES

Carbon removal includes various methods of removing carbon dioxide from the atmosphere. It consists of nature- and technology-based solutions. We have emitted so much greenhouse gas (GHG) into the atmosphere and reducing these emissions will be a challenge, carbon removal is widely considered a critical strategy to achieve our net-zero and climate goals. However, it must not distract from the urgency of drastic greenhouse gas emissions reductions.

Carbon removal also called carbon dioxide removal (CDR) will be particularly important in sectors that are especially difficult to decarbonizes, like steel, cement, and petrochemicals. "The deployment of CDR to counterbalance hard-to-abate residual emissions is unavoidable if net-zero CO_2 or GHG emissions are to be achieved," the IPCC says. For companies looking to invest in carbon removal solutions to supplement their emissions reduction efforts, it's important to understand the different types of removal, their benefits and drawbacks, and how removal financing fits in with offsets and carbon credits.

THESE SOLUTIONS RELY ON TECHNOLOGY TO REMOVE CARBON FROM THE ATMOSPHERE.
THEY INCLUDE

1 Direct air capture (DAC) uses large fans to move air through a filter to absorb CO_2.

2 Direct air carbon capture and storage (DACCS) is DAC plus storage. CO_2 is extracted from the air and stored in geological formations deep underground.

3 Bioenergy with carbon capture and storage (BECCS) involves atmospheric CO_2 absorbed by plants. The plant material (biomass) is burned to produce energy, and the CO_2 released in that process is captured and stored underground.

Each method has pros and cons. As the IPCC has stated, "There are a number of CDR methods, each with different potentials for achieving negative emissions, as well as different associated costs and side effects. They are also at differing levels of development, with some more conceptual than others." These technologies have their own environmental impacts and potential risks to people in surrounding communities.

Source: https://docs.climateinteractive.org/projects/en-roads/en/latest/images/cdr/dac.0ec5cea6bf.jpg Reference link: https://www.sustain.life/blog/problem-with-carbon-re-moval-technology

CARBON MANAGEMENT

"Management is an organized approach to gain the strategic advantages of CO_2 emissions reductions. Carbon management helps organizations stay focused on achieving their targets to reduce CO emissions and their use of fossil fuels".

Carbon management combines practical, cost-effective approaches to reducing annual greenhouse gas (GHG) emissions in ways that enhance environmental performance.Carbon management is useful for identifying useful carbon dioxide (CO_2) emissions reduction strategies for cutting back the annual emissions business report to stakeholders in their CSR reports.

There are a wide range of strategies employed in carbon management: energy efficiency, low-carbon fuel substitution, renewable energy certificates, life cycle analysis, and the use of new technologies like carbon capture, otherwise known as direct air capture – are all strategies that Carbon can help businesses lower their reported (CO_2) emissions.Science and research are seeking new ideas for decarburization technologies are being made all of the time to improve upon sustainable development, but the good news is – emissions reductions can already be made through the implementation of one or more of these broad categories. Carbon management combines practical, cost-effective approaches to reducing annual greenhouse gas (GHG) emissions in ways that enhance environmental performance.Carbon management is useful for identifying useful carbon dioxide (CO_2) emissions reduction strategies for cutting back the annual emissions business report to stakeholders in their CSR reports.

There are a wide range of strategies employed in carbon management: energy efficiency, low-carbon fuel substitution, renewable energy certificates, life cycle analysis, and the use of new technologies like carbon capture, otherwise known as direct air capture – are all strategies that can help businesses lower their reported (CO_2) emissions. Science and research are seeking new ideas for decarburization technologies are being made all of the time to improve upon sustainable development, but the good news is – emissions reductions can already be made through the implementation of one or more of these broad. categories.

Reference link: https://greenly.earth/en-gb/blog/company-guide/what-is-carbon-management

CARBON CALCULATOR

A carbon calculator estimates carbon footprints. It measures greenhouse gas emissions for a snapshot in time. Carbon calculators are used to calculate greenhouse gas inventories of facilities or operations in order to determine the amount of greenhouse gases produced for a specified year. The results can be used to prepare plans for actions to reduce the amount of greenhouse gas emitted annually or by a target year. A Carbon Calculator is a programme that calculates your Carbon Footprint. It calculates based on the amount of carbon dioxide produced by activity. Any individual or group can get their carbon footprint.

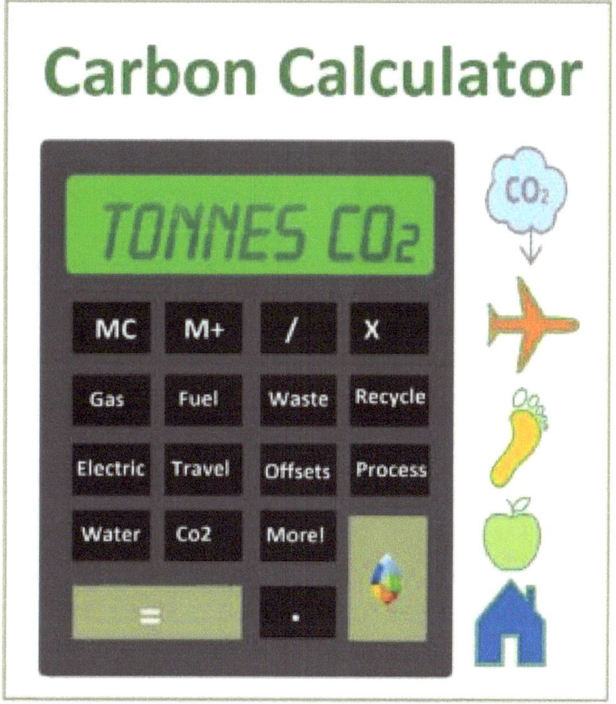

The calculator plugs the requested information into a computer programme. This information is then calculated as a carbon footprint. Information can include how we travel, heat our homes, dispose of our waste, etc.

Knowing the carbon footprint of your home or workplace makes it much easier to get involved in reducing that footprint. Before using a carbon calculator, take a few minutes to compile the information you might need. This ensures that the calculation you get will be correct.

TO CALCULATE A CARBON FOOTPRINT, QUESTIONS ARE ASKED RELATED TO THE FOLLOWING:

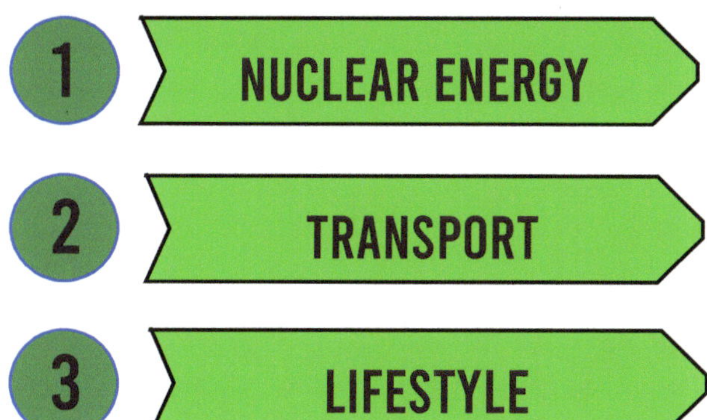

1 NUCLEAR ENERGY

2 TRANSPORT

3 LIFESTYLE

Source: https://occupli.com/carbon-footprint-calculator/
Reference link: https://www.epa.ie/take-action/in-the-home/climate-change/carbon-footprint-calculators/#:~:text=A%20Carbon%20Calculator%20is%20a,information%20into%20a%20computer%20programme

CARBON CAPTURE AND STORAGE

Carbon capture and storage (CCS) is a way of reducing carbon dioxide (CO_2) emissions, which could be to helping to tackle global warming. It's a three-step process, involving; capturing the CO_2 produced by power generation or industrial activity, such as hydrogen production, steel or cement making; transporting it; and then permanently storing it deep underground. Here we look at the potential benefits of CCS and how it works.

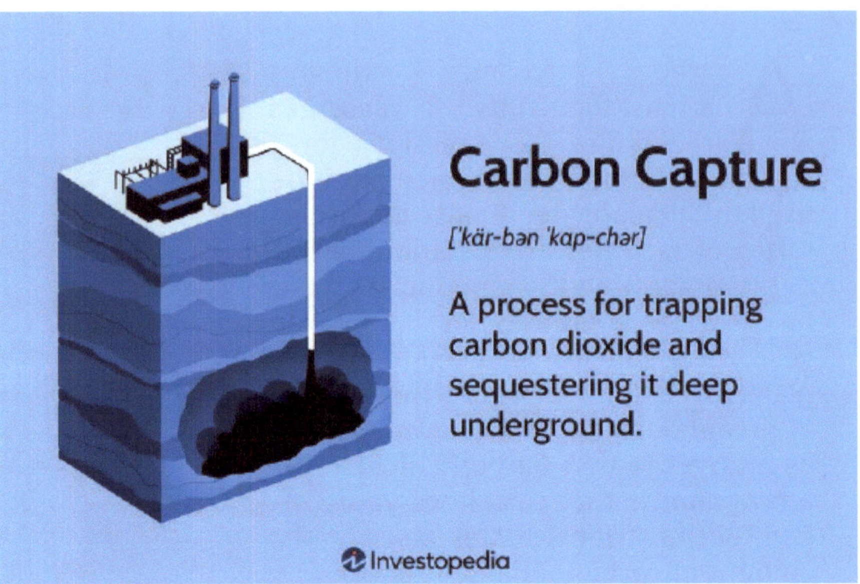

THERE ARE THREE STEPS TO THE CCS PROCESS:

1. CAPTURING THE CO_2 FOR STORAGE

The CO_2 is separated from other gases produced in industrial processes, such as those at coal and natural-gas-fired power generation plants or steel or cement factories.

2. TRANSPORT

The CO_2 is then compressed and transported via pipelines, road transport or ships to a site for storage.

3. STORAGE

Finally, the CO_2 is injected into rock formations deep underground for permanent storage.

Source: https://www.investopedia.com/carbon-capture-7973911 Reference link: https://www.nationalgrid.com/stories/energy-explained/what-is-ccs-how-does-it-work
https://en.m.wikipedia.org/wiki/Carbon_capture_and_storage

Carbon sequestration – the practice of removing carbon dioxide (CO_2) from the atmosphere and storing it – is one of the many approaches being taken to tackle climate change. Find out why this method is being used and the different ways in which CO_2 is being removed and stored.

Preventing the earth's atmosphere from warming any further is taking a huge collective effort by humanity. From ending our dependency on carbon-emitting fuels to establishing a legally binding net zero emissions target by 2050, every potential solution is important if we're to stop unprecedented climate change.

Alongside a transition to clean energy systems and decarburizing high-emission practices – such as construction or transport – humankind is making a concerted effort to remove CO_2 from our atmospheres; by adapting the ways we construct, consume, travel and generate power. But methods like carbon sequestration show how we can work with the natural environment to tackle the climate crisis.

OW DOES CARBON SEQUESTRATION WORK?

Carbon sequestration is the capturing, removal and permanent storage of CO_2 from the earth's atmosphere. It's recognized as a key method for removing carbon from the earth's atmosphere.

This is important, as around 45% of the CO_2 emitted by human's remains in the atmosphere, which is a significant factor behind global warming. Carbon sequestration can prevent further emissions from contributing to the heating of the planet.

Carbon sequestration can happen in two basic forms: biologically or geologically. Also, while it's being encouraged artificially through various biological and geological methods, it also happens naturally in the environment on the biggest scale.

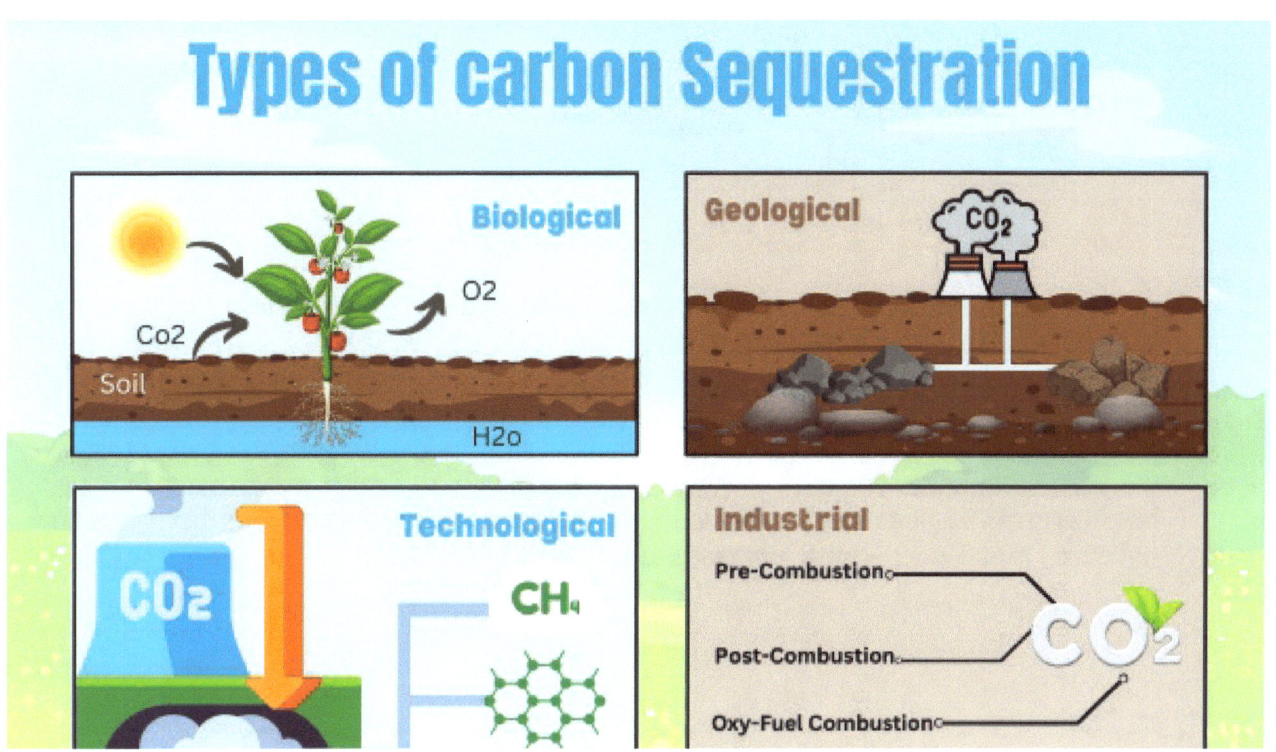

BIOLOGICAL CARBON SEQUESTRATION

Biological carbon sequestration is the storage of carbon dioxide in vegetation such as grasslands or forests, as well as in soils and oceans.

GEOLOGICAL CARBON SEQUESTRATION

Geological carbon sequestration is the process of storing carbon dioxide in underground geologic formations, or rocks. Typically, carbon dioxide is captured from an industrial source, such as steel or cement production, or an energy-related source, such as a power plant or natural gas processing facility and injected into porous rocks for long-term storage.

TECHNOLOGICAL CARBON SEQUESTRATION

Scientists are exploring new ways to remove and store carbon from the atmosphere using innovative technologies. Researchers are also starting to look beyond removal of carbon dioxide and are now looking at more ways it can be used as a resource.

INDUSTRIAL CARBON SEQUESTRATION

Industrial carbon sequestration, also known as carbon capture and storage (CCS), is a process that involves capturing carbon dioxide (CO_2) from industrial sources and storing it in a way that prevents it from entering the atmosphere. The goal of CCS is to reduce carbon dioxide emissions and help mitigate climate change.

IMPACTS OF CARBON SEQUESTRATION

About 25% of our carbon emissions have historically been captured by Earth's forests, farms and grasslands. Scientists and land managers are working to keep landscapes vegetated and soil hydrated for plants to grow and sequester carbon.

As much as 30% of the carbon dioxide we emit from burning fossils fuels is absorbed by the upper layer of the ocean. But this raises the water's acidity, and ocean acidification makes it harder for marine animals to build their shells. Scientists and the fishing industry are taking proactive steps to monitor the changes from carbon sequestration and adapt fishing practices.

urce: https://energytheory.com/carbon-sequestration-benefits/ Reference link: https://www.nationalgrid.com/stories/energy-explained/what-carbon-sequestration
https://en.m.wikipedia.org/wiki/Carbon_sequestration https://climatechange.ucdavis.edu/climate/definitions/carbon-sequestration

Carbon accounting, or greenhouse gas accounting, is the process of quantifying the number of greenhouse gases (GHGs) produced directly and indirectly from a business's or organization's activities within a set of boundaries.

Carbon dioxide (CO_2) is the most common greenhouse gas emitted by human activities. As a result, all other major GHGs are given a carbon dioxide equivalent or CO_2e. This is determined by multiplying the amount of a GHG by its global warming potential (GWP).

A gas's GWP is a measure of how much energy the emissions of 1 ton of that gas absorbs over a given period of time relative to the emissions of 1 ton of carbon dioxide. The higher the GWP, the more that GHG contributes to global warming.

WHY IS CARBON ACCOUNTING IMPORTANT?

Carbon accounting is important for at least three reasons:

ESG 1 reporting: Environmental, social, and governance (ESG) refers to a set of standards for a company's behavior. It considers how a company safeguards the environment—including corporate policies addressing climate change, for example—and is increasingly popular among socially conscious investors looking to screen potential investments. Carbon accounting provides a measure of a company's environmental impact—the "E" in "ESG"—and can therefore help it reduce risk and attract investment. Proof that companies are meeting their commitments under environmental legislation: Many countries, including the United States and the United Kingdom, now require companies to report their environmental impact (and take steps to reduce it). Carbon accounting is the standard method for this reporting.

Efficiency: More generally, carbon accounting can be important from an efficiency perspective. Goods and services that have a high carbon cost are often those that are produced inefficiently, so reducing its carbon expenditures can save a company money, as well.

CHALLENGES OF CARBON ACCOUNTING

Though carbon accounting has become a widely used tool, it can present serious challenges for organizations seeking to implement it rigorously. These include:

Significant time and expense: "Many organizations run their annual carbon accounting and ESG ratings calculation process using manual data collection and spreadsheets," IBM notes. "This leads to enhanced risk and productivity loss, especially for complex, global organizations that report to multiple frameworks."

Difficulty obtaining accurate data: For a company to accurately measure its carbon emissions, it must obtain data from many different internal and external sources, either by collecting it manually or by investing in a complex centralized system. Even if data are available, tracking the carbon impact of every purchase a company makes and every process it runs can be challenging, so errors or omissions are likely. Carbon accounts may be poorly understood at the top: Explaining the relevance and utility of carbon accounts to decision makers inside an organization can be challenging. Carbon budgets are often seen as a reporting requirement rather than a management tool in themselves.

Reference link: https://normative.io/insight/carbon-accounting-explained/ https://www.investopedia.com/carbon-accounting-7562229 https://www.ibm.com/topics/carbon-accounting#:~:-text=Carbon%20accounting%2C%20or%20greenhouse%20gas,gas%20emitte %20by%20human%20activities. https://climatechange.ucdavis.edu/climate/definitions/carbon-sequestration

CARBON BUDGET

Carbon budget is a concept used in climate policy to help set emissions reduction targets in a fair and effective way. It examines the "maximum amount of cumulative net global anthropogenic carbon dioxide (CO_2) emissions that would result in limiting global warming to a given level. It can be expressed relative to the pre-industrial period (the year 1750). In this case, it is the total carbon budget. Or it can be expressed from a recent specified date onwards. In that case it is the remaining carbon budget.Carbon budget and emission reduction scenarios needed to reach the two-degree target agreed to in the Paris Agreement (without net negative emissions, based on peak emissions)

A carbon budget that will keep global warming below a specified temperature limit is also called an emissions budget or quota, or allowable emissions.Apart from limiting the global temperature increase, another objective of such an emissions budget can be to limit sea level rise. Scientists combine estimates of various contributing factors to calculate the carbon budget. The estimates take into account the available scientific evidence as well as value judgments or choices.Global carbon budgets can be further sub-divided into national emissions budgets. This can help countries set their own emission goals. Emissions budgets indicate a finite amount of carbon dioxide that can be emitted over time, before resulting in dangerous levels of global warming. The change in global temperature is independent of the source of these emissions, and is largely independent of the timing of these emissions.

To translate global carbon budgets to the country level, a set of value judgments have to be made on how to distribute the remaining carbon budget over all the different countries. This should take into account aspects of equity and fairness between countries as well as other methodological choices there are many differences between nations, such as population size, level of industrialization, historic emissions, and mitigation capabilities. For this reason, scientists are attempting to allocate global carbon budgets among countries using various principles of equity.

Reference link: https://en.m.wikipedia.org/wiki/Carbon_budge https://carbontracker.org/carbon-budgets-explained/ https://climatechange.ucdavis.edu/climate/definitions/carbon-sequestration

CARBON TAXES

arbon tax, tax levied on firms that produce carbon dioxide (CO_2) through their operations. It is used as an incentive to reduce the economy-wide usage of high-carbon fuels and to protect the environment from the harmful effects of excessive carbon dioxide emissions. A carbon tax is levied on CO_2 emissions. All fossil fuels such as coal, petroleum, and natural gas contain carbon, which is released as carbon dioxide when these fuels are burned. The released carbon dioxide acts as a greenhouse gas: it prevents the infrared radiation generated by sunlight that has heated Earth from escaping to space efficiently, which creates a heat-trapping effect. Over time, the accumulation of greenhouse gases in the atmosphere contributes to climate change and causes nonreversible harm to the environment.

A carbon tax works on the basis of the economic principle of externalities. When a firm generates pollution through carbon dioxide emissions, it is said to produce a negative externality—a cost to the society through the harm that it causes to the environment. A carbon tax is a way to internalize that cost. In other words, it is a market-based solution that is grounded on the principle that emissions will be reduced when businesses are obliged to pay at least part of the cost of the externality they have created. Furthermore, such a tax has the potential to encourage firms to invest in environmentally friendly renewable energy and reduce the economy-wide reliance on fossil fuels. A carbon tax is easy to implement because it is based on CO_2 emissions, which is straightforward to measure, and it offers a potentially cost-effective way of reducing carbon-dioxide emissions and fossil-fuel usage. In the early 21st century, a number of countries, such as Canada, Ireland, and Sweden, began using a carbon-tax system in which firms are obligated to pay a tax based on the carbon content of the fuels they use in their production. Countries in the European Union, on the other hand, chose to partly rely on a market exchange system called the European Union Emissions Trading Scheme (ETS), where firms were allowed to buy and sell emission rights between each other. Many Organizations for Economic Co-operation and Development (OECD) and eastern European countries indirectly taxed carbon dioxide emissions through taxes on energy products and motor vehicles.

Source: https://www.britannica.com/money/carbon-tax#:~:text=carbon%20tax%2C%20tax%20levied%20on,of%20excessive%20carbon%20dioxide%20emissions.
Reference link: https://en.m.wikipedia.org/wiki/Carbon_tax https://u4d2z7k9.rocketcdn.me/wp-content/uploads/2020/01/Webp.net-resizeimage-27-1024x683.jpg

CARBON AUDITING

A carbon audit, also known as a carbon footprint assessment, is essentially a systematic examination of the greenhouse gas (GHG) emissions produced by an individual, organization, building, or event. It's like a financial audit for your environmental impact, giving you a clear picture of where your emissions are coming from and how much you're generating.

WHY IS A CARBON AUDITING IMPORTANT?

Businesses and organizations take on many initiatives to employ sustainable sources of energy and methods to improve efficiency and encourage better carbon management. To measure these efforts and compare the impact, incorporating accounting standards for emissions becomes crucial. A carbon audit is a scientific method to quantify carbon emissions and, and impact and can help the company make effective carbon management.

WHAT ARE THE BENEFITS OF CARBON AUDIT FOR THE BUSINESS?

When the business makes active efforts to reduce carbon emissions, it is not only good for the environment but also paves the way to multiple business and investment opportunities.

Eco-Efficiency	: Pinpoint energy waste for cost-effective resource use.
Regulatory Assurance	: Ensure compliance with environmental standards and regulations.
Reputation Boost	: Enhance brand image through visible sustainability efforts.
Risk Mitigation	: Identify and manage climate-related risks for resilience.
CostSavings	: Optimize operations, lowering utility expenses and inefficiencies.
Employee Engagement	: Foster staff pride and engagement in eco-friendly practices.
Investor Appeal	: Attract socially responsible investors with green initiatives.
Innovation Opportunities	: Inspire innovation by identifying efficiency improvements.
SupplyChain Sustainability	: Evaluate and enhance the sustainability of the entire supply chain.
Stakeholder Trust	: Address environmental concerns transparently, meeting expectations.
Long-TermResilience	: Position for success in a market valuing sustainable practices.

SCOPE OF THE CARBON AUDIT

The scope of the carbon audit is defined by the activities that are carried out for, and invoiced according to the project actions. Additionally, the emissions from the supply chain are also included. Eligible companies must follow the guidelines mentioned by the Department of Environment and Rural Affairs (Defra) for UK organizations and businesses. The qualified companies are those who are complying with the GHG regulations

Reference link: https://imveloltd.co.uk/carbon-audit-a-practical-guide-for-businesses/

CARBON LEAKAGE

"CARBON LEAKAGE IS DEFINED AS THE INCREASE IN CO_2 EMISSIONS OUTSIDE OF COUNTRIES THAT ARE TAKING DOMESTIC MITIGATION ACTION DIVIDED BY THE REDUCTION IN EMISSIONS FROM THESE COUNTRIES."

It is given as a percentage and might be larger or less than 100%.Changes in trading patterns can cause carbon leakage, which is frequently assessed as the balance of emissions embodied in trade (BEET).Carbon leakage happens when greenhouse gas emissions in one nation increase as a result of emissions reductions in a second country with rigorous climate policies. The transfer of carbon-intensive firms out of industrialized economies is referred to as carbon leakage

This may result in an increase in their overall emissions. Certain energy-intensive industries may be more vulnerable to carbon leakage. One sort of spill-over impact is carbon leakage. Spill-over effects can be good or bad; for example, emission reduction policies may result in technical advances that help in emission reductions outside of the policy region.

This implies that the domestic strategy for mitigating climate change is less efficient and more expensive in limiting emission levels, which is a real worry for policy-makers. Carbon Tax, also known as Carbon Cass, is one example of the strictest regulations that is imposed in several nations.

REASONS OF CARBON LEAKAGE

Carbon leakage can occur for a variety of reasons, including:

> If a country's emissions policy boosts local costs, another country with a more lenient policy may have a trading advantage.

> If demand for these commodities remains constant, production may shift offshore to a cheaper nation with lesser regulations, with no reduction in world emissions.

> If one country's environmental regulations raise the price of particular fuels or commodities, demand will fall and the price will fall.

> Countries that do not charge a premium for certain commodities may fill the demand and use the same supply, negating any gain.

> There is no agreement on the size of the long-term leaking consequences. This is critical for the issue of climate change.

Reference link: https://en.m.wikipedia.org/wiki/Carbon_leakage https://prepp.in/news/e-492-carbon-leakage-environment-notes

CARBON PRICING
(DIRECT OR EXPLICIT CARBON TAX / ETS)

Carbon pricing is an instrument that captures the external costs of greenhouse gas (GHG) emissions the costs of emissions that the public pays for, such as damage to crops, health care costs from heat waves and droughts, and loss of property from flooding and sea level rise and ties them to their sources through a price, usually in the form of a price on the carbon dioxide (CO_2) emitted. A price on carbon helps shift the burden for the damage from GHG emissions back to those who are responsible for it and who can avoid it. Instead of dictating who should reduce emissions where and how, a carbon price provides an economic signal to emitters, and allows them to decide to either transform their activities and lower their emissions, or continue emitting and paying for their emissions. In this way, the overall environmental goal is achieved in the most flexible and least-cost way to society. Placing an adequate price on GHG emissions is of fundamental relevance to internalize the external cost of climate change in the broadest possible range of economic decision making and in setting economic incentives for clean development. It can help to mobilize the financial investments required to stimulate clean technology and market innovation, fueling new, low-carbon drivers of economic growth.

There is a growing consensus among both governments and businesses on the fundamental role of carbon pricing in the transition to a decarbonized economy. For governments, carbon pricing is one of the instruments of the climate policy package needed to reduce emissions. Finally, long-term investors use carbon pricing to analyze the potential impact of climate change policies on their investment portfolios, allowing them to reassess investment strategies and reallocate capital toward low-carbon or climate-resilient activities.

Carbon pricing can take different forms and shapes. In general, both the State and Trends of Carbon Pricing series and the Carbon Pricing Dashboard focus on direct carbon pricing instruments – that is, those that apply a price incentive directly proportional to the greenhouse gas emissions generated by a given product or activity (primarily carbon taxes, ETSs, and carbon crediting mechanisms). While these policies are called "carbon taxes" and have been historically included within the State and Trends Reports, they are closer to the definition for indirect carbon pricing, due to non-uniform carbon prices across fuels

Reference link: https://carbonpricingdashboard.worldbank.org/what-carbon-pricing

(INDIRECT OR IMPLICIT- FUEL TAX/REDUCTION IN FUEL SUBSIDY/OTHERS)

Fuel Subsidies are intended to protect consumers by keeping prices low, but they come at a substantial cost. Subsidies have sizable fiscal consequences (leading to higher taxes/borrowing or lower spending), promote inefficient allocation of an economy's resources (hindering growth), encourage pollution (contributing to climate change and premature deaths from local air pollution), and are not well targeted at the poor (mostly benefiting higher income households). Removing subsidies and using the revenue gain for better targeted social spending, reductions in inefficient taxes, and productive investments can promote sustainable and equitable outcomes. Fossil fuel subsidy removal would also reduce energy security concerns related to volatile fossil fuel supplies.

Subsidies are decomposed into explicit and implicit subsidies. Explicit subsidies occur when the retail price is below a fuel's supply cost. For a non-tradable product (e.g., electricity), the supply cost is the domestic production cost, inclusive of any costs to deliver the energy to the consumer, such as distribution costs and margins. In contrast, for an internationally tradable product (e.g., oil), the supply cost is the opportunity cost of consuming the product domestically rather than selling it abroad plus any costs to deliver the energy to the consumer. Explicit subsidies also include direct support to producers, such as accelerated depreciation, but these are relatively small.

Implicit subsidies occur when the retail price fails to include external costs, inclusive of the standard consumption tax. External costs include contributions to climate change through greenhouse gas emissions, local health damages (primarily pre-mature deaths) through the release of harmful local pollutants like fine particulates, and traffic congestion and accident externalities associated with the use of road fuels. Getting energy prices right involves reflecting these adverse effects on society in prices and applying general consumption taxes when fuels are consumed by household.

CARBON PRICING POLICIES

The fight against climate change requires a multifaceted approach, where carbon pricing serves as a critical component within a broader suite of climate policies. To maximize the effectiveness of carbon pricing and ensure a holistic response to climate change, it is essential to integrate it with complementary policies. These policies can address areas not fully covered by carbon pricing and help to mitigate potential negative impacts, ensuring a just and equitable transition to a low-carbon economy.

RENEWABLE ENERGY INCENTIVES:

While carbon pricing encourages the reduction of greenhouse gas emissions by making fossil fuels more expensive, renewable energy incentives such as tax credits, grants, and feed-in tariffs directly support the development and deployment of clean energy technologies. These incentives can accelerate the transition to renewable energy by making it more financially attractive for businesses and consumers, thereby reducing reliance on fossil fuels.

ENERGY EFFICIENCY STANDARDS:

Implementing energy efficiency standards for buildings, appliances, and vehicles can significantly reduce energy demand and emissions. Carbon pricing alone may not be sufficient to drive the adoption of energy-efficient technologies and practices, especially in sectors where the cost of switching to more efficient options is high. Standards and regulations can ensure that energy efficiency improvements are implemented across the board, complementing the price signals sent by carbon pricing.

RESEARCH AND DEVELOPMENT (R&D) FUNDING:

Investing in R&D for new technologies is crucial for long-term decarbonization. Carbon pricing generates revenue that can be allocated to support R&D in areas such as carbon capture and storage (CCS), advanced renewable energy technologies, and battery storage solutions. This investment can lead to breakthroughs that reduce the cost of clean technologies and make them more accessible.

ADAPTATION AND RESILIENCE BUILDING:

Climate policies must also address adaptation to the impacts of climate change, particularly in vulnerable communities. Funds raised through carbon pricing can support infrastructure improvements, ecosystem restoration, and community-based adaptation projects. These efforts can help societies prepare for and respond to the effects of climate change, reducing their vulnerability and enhancing resilience.

SOCIAL EQUITY PROGRAMS:

To address concerns about the regressive nature of carbon pricing and its potential to disproportionately affect low-income households, revenues can be used to fund social equity programs. These programs can include direct rebates to households, investments in public transportation, and support for energy efficiency upgrades in low-income communities. By doing so, carbon pricing can contribute to social equity and reduce the burden on those least able to afford the transition to a low-carbon economy.

Reference link: https://www.linkedin.com/pulse/carbon-pricing-cornerstone-climate-mitigation- sustainable-sahil-baxi-dtpvf

Carbon Pricing Metrics (nECR/TCP)

Total Carbon Pricing (TCP) is to provide a comprehensive picture of the extent to which economiesprice the social cost of GHG emissions across countries, sectors, and fuels. The metric willincrease the informational base for the academic and policy discussions surrounding carbon pricing comparability, minimum carbon price commitments, and rules to understand the incentivesunder future carbon border adjustment mechanisms. To achieve these aims, the TCP needs toreflect not only prices or tax rates but also effective coverage, because many carbon pricing systemsonly cover emissions from specific energy-intensive industries 2 or installations above minimumemissions thresholds. The TCP should also reflect effective rates, net of special exemptions 3 andreduced rates, because special provisions for specific sectors may erode the price signal. Finally,the TCP should be transparent and easily verifiable, requiring using publicly available data andminimizing modeling, particularly where assumptions are not easily verifiable.

CARBON PRICING INSTRUMENTS

A policy vehicle, implemented through a legal and institutionalinfrastructure that can deliver a price on carbon emissions on specific sectors and/or entities.

CARBON TAX:

A carbon tax directly sets a price on carbon by defining a tax rate on greenhouse gas emissions or the carbon content of fossil fuels. It is straightforward to administer and provides a predictable price on carbon, making it easier for businesses to plan for and invest in low-carbon technologies.

ADVANTAGES:

Simplicity and transparency make it easier to implement and understand. Provides a stable environment for businesses to make long-term investments in reducing emissions. Revenue generated from carbon taxes can be used to fund renewable energy projects, lower other taxes, or be redistributed to mitigate the economic impact on consumers.

CHALLENGES:

Determining the optimal tax rate can be complex and politically sensitive. Without proper design, it may not guarantee a specific level of emissions reduction. Risk of economic burden on low-income households if not implemented with appropriate compensatory measures.

Reference link: https://documents1.worldbank.org/curated/
en/099548206152339098/pdf/IDU124d2b624145531468a1a4d418173bf51a4fd.pdf

EMISSIONS TRADING SYSTEMS (ETS):

Also known as cap-and-trade systems, ETS sets a cap on the total level of greenhouse gas emissions and allows industries with low emissions to sell their extra allowances to larger emitters. This creates a market price for carbon emissions and incentivizes companies to reduce their emissions to sell their allowances.

ADVANTAGES:

Guarantees a certain level of emissions reduction by setting a cap on total emissions. Encourages cost-effective emissions reductions where they are most economical. Can be linked with other ETS globally to increase market flexibility and liquidity.

CHALLENGES:

Requires a robust monitoring and reporting system to ensure compliance. The price of carbon can be volatile, making it harder for businesses to predict costs. Allocation of allowances can be contentious and subject to lobbying.

Source: https://medium.com/@esgmcgill/carbon-pricing-instruments-an-overview-of-the-carbon-tax-and-cap-and-trade-system-6f17f0198aa2
Reference link: https://www.linkedin.com/pulse/carbon-pricing-cornerstone-climate-mitigation-sustainable-sahil-baxi-dtpvf

CARBON PRICING AS A FISCAL TOOL

Carbon pricing is a fiscal tool used by governments to put a cost on carbon emissions, providing a financial incentive for companies and individuals to reduce their greenhouse gas emissions.

There are two main types of carbon pricing:
1. Carbon Tax: A direct tax on fossil fuels, such as coal, oil, and gas, based on their carbon content.
2. Cap-and-Trade System: A market-based mechanism that sets a limit on total emissions and allows companies to buy and sell emission allowances.

Benefits of carbon pricing as a fiscal tool:
1. Reduces greenhouse gas emissions: By putting a cost on carbon, companies and individuals are incentivized to switch to cleaner energy sources and reduce their emissions.
2. Generates revenue: Carbon pricing can generate significant revenue for governments, which can be used to fund low-carbon initiatives, support vulnerable communities, or reduce other taxes.
3. Encourages sustainable behaviors: Carbon pricing can influence consumer behavior, encouraging individuals to make more sustainable choices, such as using public transport or carpooling.

Examples of carbon pricing in action:
1. Sweden: Sweden has had a carbon tax since 1991, which has helped reduce the country's greenhouse gas emissions by 23% since 1990.
2. British Columbia, Canada: British Columbia introduced a carbon tax in 2008, which has helped reduce the province's greenhouse gas emissions by 10% since 2008.
3. European Union: The European Union has a cap-and-trade system, known as the EU Emissions Trading System (EU ETS), which covers more than 11,000 power stations and industrial plants across the EU.

Challenges and limitations of carbon pricing:
1. Competitiveness concerns: Carbon pricing can increase costs for companies, potentially making them less competitive in the global market.
2. Regressive impacts: Carbon pricing can disproportionately affect low-income households, who may spend a larger portion of their income on energy and transportation.
3. Revenue allocation: The allocation of revenue generated from carbon pricing can be a challenge, with different stakeholders having competing demands for the revenue.

Overall, carbon pricing is a valuable fiscal tool for reducing greenhouse gas emissions and generating revenue for low-carbon initiatives. However, its design and implementation require careful consideration of competitiveness concerns, regressive impacts, and revenue allocation.

Reference link: https://www.imf.org/en/Publications/fandd/issues/Series/Back-to-Basics/Fiscal-Policy
Reference link : https://en.wikipedia.org/wiki/Verified_Carbon_Standard https://worldbank.scene7.com/is/image/worldbankprod/carbon-pricing-buildings:1440x600

CARBON STANDARD

The Verified Carbon Standard (VCS), formerly the Voluntary Carbon Standard, is a standard for certifying carbon credits to offset emissions. VCS is administered by Verra, a 501organization. Verra is the world's biggest certifier of voluntary carbon offsets. As of 2020 there were over 1,500 certified VCS projects covering energy, transport, waste, forestry, and other sectors. In 2021 Verra issued 300 MtCO2e worth of offset credits for 110 projects. There are also specific methodologies for REDD+ projects. Verra is the program of choice for most of the forest credits in the voluntary market, and almost all REDD+ projects.

Verra was developed in 2005 when the company Climate Wedge and its partner Cheyne Capital designed and drafted the first version (version 1.0) of the Voluntary Carbon Standard. This standard was intended as a quality standard for transacting and developing "non-Kyoto" Protocol carbon credits. Climate Wedge was at the time active as a carbon markets investment advisory firm.

CARBON BOARDER ADJUSTMENTS MECHANISM

The European Union (EU) enforcing its new Carbon Border Adjustment Mechanism (CBAM) designed to equalize the price of carbon between domestically produced goods and imports. As the EU raises the bar for its own climate ambitions, it sees also the obligation to put a fair price on the carbon emitted during the production of carbon intensive goods entering and leaving the EU.

The adjustment mechanism (CBAM) aims to enhance the European Union's environmental commitments and promote cleaner industrial practices in non-EU countries. Presently, CBAM primarily focuses on imports of cement, iron, steel, aluminum, fertilizers, electricity, and hydrogen. However, it is anticipated that the scope of its application will soon encompass a broader range of industries and products.

The scheme will be gradually introduced in several phases beginning October 1st, 2023, leading up to January 1st, 2026, at which point importers will be obligated to pay tariffs, or carbon pricing based on the prevailing prices of EU emission trading scheme allowances. This phased implementation is designed to provide a smooth transition and enable businesses to adapt to the new regulations effectively.

CARBON REMOVAL PROJECTS

EMISSION REDUCTIONS:

Carbon pricing mechanisms, such as carbon taxes and emissions trading systems (ETS), incentivize businesses and individuals to reduce their carbon footprint by making greenhouse gas emissions a costly activity. This financial motivation drives the adoption of cleaner technologies and practices, leading to a significant reduction in emissions. For instance, the European Union Emissions Trading System (EU ETS) has been instrumental in reducing emissions from covered sectors, showcasing carbon pricing's effectiveness in mitigating climate change.

INNOVATION AND CLEAN TECHNOLOGY:

By putting a price on carbon, these mechanisms also spur innovation in clean technology. The additional costs associated with emitting carbon dioxide encourage companies to invest in research and development of new, less carbon-intensive ways of operating. This innovation not only helps reduce emissions but also advances the technology needed to transition to a low-carbon economy.

BEHAVIORAL CHANGE:

Beyond the corporate world, carbon pricing can influence individual behavior, encouraging more sustainable choices such as using public transportation, reducing energy consumption, and supporting renewable energy sources. These changes in behavior are crucial for achieving the broader climate action goals outlined in SDG 13.

FUNDING FOR CLIMATE PROJECTS:

Revenue generated from carbon pricing can be allocated to climate adaptation and mitigation projects, further contributing to the goals of SDG 13. For example, funds can be used to support renewable energy projects, forest conservation, and resilience-building initiatives in vulnerable communities, directly linking carbon pricing efforts with climate action objectives.

GLOBAL LEADERSHIP AND COOPERATION:

Carbon pricing also plays a role in fostering international cooperation on climate change, a key aspect of SDG 13. By implementing carbon pricing mechanisms, countries demonstrate their commitment to global climate action, encouraging others to follow suit and contributing to the collective effort required to address climate change effectively.

Reference link: https://www.linkedin.com/pulse/carbon-pricing-cornerstone-climate-mitigation-sustainable-sahil-baxi-dtpvf

CARBON CREDIT ISSUANCE

Carbon credits are a transparent, measurable and results-based way for companies to support activities, such as protecting and restoring irrecoverable natural carbon sinks, like forests or marine ecosystems and scaling nascent carbon removal technology, that keep global climate goals within reach.

WHAT ARE CARBON CREDITS

Carbon credits are permits that allow the owner to emit a certain amount of carbon dioxide or other greenhouse gases (GHGs). One credit allows the emission of one ton carbon dioxide or the equivalent of other greenhouse gases. Carbon credits are also known as carbon offsets.

The carbon credit is half of a cap-and-trade program. Companies that pollute are issued credits that allow them to continue to pollute up to a certain limit that's periodically reduced. The company can sell any unneeded credits to other companies that need them so private companies are doubly incentivized to reduce greenhouse emissions.

Companies receive a set number of credits that decline over time. They can sell any excess credits to another company.

Carbon credits create a monetary incentive for companies to reduce their carbon emissions.

Carbon credits are based on the cap-and-trade model that was used to reduce sulfur pollution in the 1990s.

Negotiators at the Glasgow COP26 climate change summit agreed in November 2021 to create a global carbon credit offset trading market.

TYPES OF OFFSET PROJECTS

- RENEWABLE ENERGY
- METHANE COLLECTION AND COMBUSTION
- ENERGY EFFICIENCY
- DESTRUCTION OF INDUSTRIAL POLLUTANTS

CARBON CREDIT RETIREMENT

A carbon credit is a reduction in greenhouse gas emission to compensate for emissions made somewhere else. A carbon credit is retired once its benefit has taken place.

THE RETIREMENT PROCESS EXPLAINED

A carbon credit is retired once its benefit has taken place. That means it has been used and the carbon benefit it represents has been claimed by the entity that bought I Retiring your carbon credits requires you to ensure that they are removed from the marketplace and labeled as 'retired' in any records or registry. The retired credits must serve their emission reduction purpose only once to prevent double counting. Take note that retirement only occurs once the impact has happened. This means retiring your carbon credits depends on what type of credit you purchase.   If you've bought ex-post carbon credits, you can retire them right after your purchase. You can then instantly get the proof of retirement. For ex-ante and pre-purchase carbon credits, retiring them won't happen immediately after you bought them. That's because their impact hasn't yet occurred and their retirement should be in the future. You should know when the timeline would be from the seller or the marketplace where you purchase the credits. It may take months or even years depending on the specific project you invest in.

THE LIFECYCLE OF A CARBON CREDIT

These initiatives often yield positive benefits to the environment and local communities. More importantly, each credit retired helps quantify the actual environmental impact of those projects. When it comes to the impact of retiring carbon credits on investors, be it individuals or companies, it has two major effects. First, it preserves the integrity and effectiveness of emission reduction projects. It prevents double counting or reusing of the credits by multiple entities. This further guarantee transparency and accountability in the carbon markets. In effect, carbon credit retirement instils confidence among companies regarding the impact of their purchases or investments.

Reference link: https://www.investopedia.com/terms/c/carbon_credit.asp https://en.wikipedia.org/wiki/Carbon_offsets_and_creditshttps://carboncredits.com/retiring-carbon-credits-everything-you-need-to-know/

CARBON CREDIT INSURANCE

Carbon credit insurance solutions support the growth of the global carbon markets. Insurance can mitigate financial risks for buyers and sellers of carbon credits. Contact Us. Carbon markets can accelerate action to combat climate change

WHY INSURE?

Third-party guarantees are a way for project credits to be CORSIA eligible and American Carbon Registry, Gold Standard and Verra all state insurance as guarantee. The insurance solution, Corresponding Adjustment Protect is a pathway to ensure market access.

INSURANCE SOLUTIONS FOR THE CARBON MARKET

1.NON-DELIVERY RISK:

Cover for the pre-payment amount against later non-delivery of the carbon credits.

2.POLITICAL RISK INSURANCE:

Protection against the financial loss of carbon credits   resulting from political perils.

3.NON-PAYMENT INSURANCE FOR BANKS AND FINANCIAL INSTITUTIONS:

Examples of insurable transactions to mitigate against non-payment risks specific to carbon credits:

1. Pre-payment/ open-account facilities
2. Borrowing base

CARBON OFFSET INSURANCE

The product protects clients against loss due to invalidation of the Carbon Offsets they hold on their balance sheet. Should covered Offsets be invalidated by a relevant authority the product provides the funds to purchase replacement Offsets.

ROLE OF INSURANCE IN CARBON MARKET

The product protects clients against loss due to invalidation of the Carbon Offsets they hold on their balance sheet. Should covered Offsets be invalidated by a relevant authority the product provides the funds to purchase replacement Offsets.

ACCORDING TO THE REPORT, INSURANCE CAN PROVIDE 4 KEY BENEFITS TO THE CARBON MARKET

A balance between traditional risk management practices and innovation – enabling improved access to finance to scale carbon projects.

A stamp of confidence – risk management and regulatory expertise, honed over decades, can bring confidence to the market and its participants.

Detailed assessment of carbon project risk – highlighting areas of concern across the market and project types, where wider risk management improvements are required.

Encourage market participants to take risks – insurers take on responsibility when things go wrong, giving market actors the freedom to take risks which are necessary to release capital and scale carbon projects and their associated benefits.

HOW THE CARBON INSURANCE WORK

Carbon credit insurance protects your buyers from unforeseen losses and unavoidable risks. Our AI-powered underwriting platform aggregates multiple data sources to generate a comprehensive quote. We provide financial compensation for carbon credit losses and resolve claims quickly.

UNLOCK THE MARKET

Carbon Insurance is the essential guarantee that protects your buyers in a landscape of changing risks to achieve net zero.Our carbon insurance products, Corresponding Adjustment Protect™ and Carbon Protect™, provide your buyers with financial compensation in the event of losses outside of your control.

1. Carbon credit insurance protects your buyers from unforeseen losses and unavoidable risks.

2.Our AI-powered underwriting platform aggregates multiple data sources to generate a comprehensive quote.

3.We provide financial compensation for carbon credit losses and resolve claims quickly.

Reference link: https://carboninsurance.co/

CARBON INSURANCE COMPANY

IN ESSENCE, CARBON INSURANCE COMPANIES HELP:

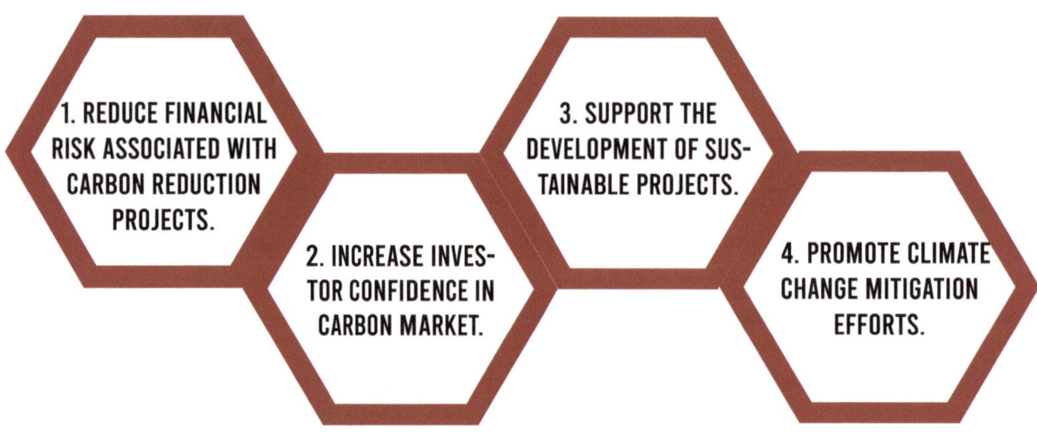

TOP 05 CARBON CREDITS COMPANIES AND STARTUPS IN INDIA IN AUGUST 2024 CARBON CREDITS COMPANIES' SNAPSHOT

They're tracking and more Carbon Credits companies in India from the F6S community. Carbon Credits forms part of the energyindustry, which is the 16th most popular industry and market group. If you're interested in the Energy market, also check out the top energy and Cleantech, Renewable Energy,energy efficiency and recycle or companies.

1.CARBO CAP

Carbocap provides carbon credit project developers with a blockchain based solutions for granular data capture for credit provenance and project development. It also provides buyers a secure marketplace for buy/hold/sell/verification of said credits.

2.SAMAK TECHNOLOGIES (BRAND:GAU RAKSHAK)

Gau Rakshak™ is one of the fastest growing start-ups in Dairy Tech sector and one of the very few companies providing end-to-end value chain solutions and services to the dairy farming community in India. Our Vision is to Bolster farmers in delivering sustainable on-farm practices through mutual involvement of all ecosystem stakeholders and promoting green technology.

3.EQUATOR GEO PRIVATE LIMITED

Currently, climate change mitigation efforts in India rely on National Action Plan for Climate Change (NAPCC) and State Action Plan for Climate Change (SAPCC). The government has committed to be a party in reducing global warming to 1.5 degree Celsius in the Paris Agreement.

4.CARBON FIXERS

'Carbon fixers' makes businesses and infrastructure Climate Resilient and Sustainable through Strategies, Technologies and Resource Efficiency. We are technology aggregators, business model innovators and system integrators to reduce your carbon footprint. Sustainability is a strategy that helps an organization manage their resources more efficiently and responsibly.

5. RE-PURPOSE GLOBAL

Re-Purpose is a global movement of conscious consumers and businesses going Plastic Neutral. As the world's first Plastic Credit Platform, we make climate action delightfully simple for companies of all sizes by removing and recycling as much ocean-bound plastic waste as theyproduce while embedding sustainability into their product experience to reach and retain purposeful consumers.

Reference link: https://www.f6s.com/companies/carbon-credits/india/co

CARBON MITIGATION

MITIGATION

Reducing emissions of and stabilizing the levels of heat-trapping greenhouse gases in the atmosphere ("mitigation"); Adapting to the climate change already in the pipeline ("adaptation").

MITIGATION CARBON CAPTURE

What is carbon capture, usage and storage (CCUS)? - CCUS refers to a suite of technologies that enable the mitigation of carbon dioxide (CO_2) emissions from large point sources such as power plants, refineries and other industrial facilities, or the removal of existing CO_2 from the atmosphere.

CARBON FOOTPRINT MITIGATION

NASA defines carbon mitigation as the concerted effort to diminish and stabilize levels of heat-trapping greenhouse gases in the atmosphere. Adaptation, on the other hand, involves adjusting life to the current and anticipated impacts of climate change.

A MINDSET FOR CARBON MITIGATION

1. Understand the unique carbon footprint of your industry. Different sectors contribute to emissions in diverse ways. Net0's dashboard helps businesses see their carbon footprint and their progress mapped out. This gives companies a qualitative and quantitative big picture of their emissions.

2. Recognize regional variations in climate impact and vulnerability. Adapt the carbon mitigation strategy to address the specific climate-related risks and opportunities associated with the geographical location of your operations.

3.Assess the resources at your disposal, considering both financial and technological aspects. Determine the feasibility of implementing various carbon mitigation measures based on the available resources. Use a combination of AI and a green mindset to focus on your carbon reduction strategy. With Net0's marginal abatement cost curve tool, businesses can easily see what carbon reduction scenarios would be cost-effective for them at the moment and what they can plan on down the line.

4. Stay informed about local and global regulations related to carbon emissions. Compliance with should be at the front end of   yourcarbon mitigation strategy to avoid legal issues and capitalize on incentives. Net is compliant with all global regulations, obligatory or voluntary.

5. Involve key stakeholders, including employees, customers, investors, and local communities. Garner support and input to ensure the success of the mitigation strategy, and communicate transparently about your goals and progress.

eference link: https://net0.com/blog/carbon-mitigation

CARBON EMISSION AVOIDANCE PROJECTS

CARBON AVOIDANCE PROJECTS

Carbon avoidance projects are proactive measures that halt additional greenhouse gases from entering the atmosphere. By replacing high-emission sources like fossil fuel plants with renewable energy alternatives, such as wind or solar farms, these projects ensure that potential emissions are never released.

1.KHASI HILLS COMMUNITY REDD+ PROJECT

Through 2022, the Khasi project has paid $472,984 to approximately 7,764 families and has invested $298,817 in forest conservation and management. Since 2011, the project has been protecting and restoring 23,512 hectares of forests, and has verifiably avoided over 411K tonnes of CO_2 emissions.

OVERVIEW

The Khasi Hills Community REDD+ Project is located in the remote northeast state of Meghalaya with Bangladesh to the south and Bhutan to the north. Spanning 23,512 hectares, it deploys strategies for both forest protection (Reducing Emissions from Deforestation and Forest Degradation, or "REDD") and restoration (Assisted Natural Regeneration, or "ANR"). The project also provides detailed and long-term plans for improving the livelihoods of 4,400 households, 80 to 90% of which live below the poverty line. The region experienced a rapid 28% loss in forest cover between 2000 and 2005. Further, India ranks 136th of 186 countries in the UNDP's 2012, and annual incomes in the region are $490 for a family of 5 to 6.

2. DRAWA RAINFOREST CONVERSATION FOREST

Since 2012, the Drawa Rainforest Conservation project has created $283,063 in income for approximately 120 inland village households spanning 1,588 hectares across the Cakadrove and Macuata provinces of Vanua Levu, Fiji. Drawa is expected to remove and avoid 436K tonnes of CO_2 emissions over 30 years, 76K of which have been verified.

3.BUJANG RABA COMMUNITY REDD+ CARBON PROJECT

Since 2014, the project has generated $362,845 for 934 households in the tropical mountainous forests of Bukit Panjang Rantau Bayur,where household incomes are $4-5/day. It has resulted in the avoidance of 303K tonnes of CO_2 emissions by protecting 5,339 hectares on the critical frontier of Sumatra's last intact rainforests.

Through 2022, Trees for Global Benefits has paid out $4.76 million to 26,555 participating households spanning 18,140 hectares and 19 districts. Trees planted by the project through 2022 are projected to sequester 3.89 million tonnes of CO_2.

OVERVIEW

Eco trust's Trees for Global Benefits (TFGB) project, located in within the Albertine Rift valley in Uganda, is a small-scale, farmer-led agroforestry program which produces long-term, Plan Vivo-accredited emissions reductions while measurably improving farmer livelihoods and emphasizing sustainable land-use practices. Uganda, in sub-Saharan East Africa, is amongst the poorest countries in the world, ranked 163th of 187 countries in the UNDP's 2014. The TFGB project began as a pilot in the Bushenyi district in 2003, and the pilot's success has led to expansion into the districts of Rubirizi, Mitooma, Kasese, Hoima, and Masindi, as well as into the Mount Elgon ecosystems of Rwenzori, Mbale, Manafwa, Bududa, Bulambuli, and Sironko.

CARBON AVOIDANCE: PREVENTING EMISSIONS BEFORE THEY OCCUR

Carbon avoidance refers to the actions taken to prevent greenhouse gas emissions from occurring in the first place. This approach is often more cost-effective and efficient than trying to reduce or offset emissions after they have occurred.

Types of carbon avoidance:

1. Energy Efficiency: Improving the energy efficiency of buildings, appliances, and industrial processes to reduce energy consumption.
2. Renewable Energy: Transitioning to renewable energy sources like solar, wind, and hydroelectric power to reduce dependence on fossil fuels.
3. Sustainable Land Use: Implementing sustainable agriculture practices, reforestation, and conservation efforts to sequester carbon dioxide.
4. Transportation Electrification: Promoting the adoption of electric vehicles and public transportation to reduce emissions from transportation.
5. Waste Reduction and Recycling: Implementing waste reduction, recycling, and composting programs to minimize waste-related emissions.

Benefits of carbon avoidance:

1. Cost Savings: Avoiding emissions can be more cost-effective than reducing or offsetting them.
2. Increased Energy Security: Reducing dependence on fossil fuels can improve energy security and reduce the risks associated with price volatility.
3. Job Creation and Economic Growth: Investing in carbon avoidance technologies and practices can create new job opportunities and drive economic growth.
4. Improved Public Health: Reducing air pollution from fossil fuels can improve public health and reduce the economic burdens associated with air pollution-related illnesses.
5. Enhanced Environmental Protection: Carbon avoidance can help protect ecosystems, preserve biodiversity, and maintain ecosystem services.

To implement carbon avoidance strategies, governments, businesses, and individuals can:

1. Set ambitious emission reduction targets
2. Invest in research and development of new technologies
3. Implement policies and regulations that support carbon avoidance
4. Promote education and awareness about the importance of carbon avoidance
5. Encourage behavioral changes and sustainable lifestyle choices

Reference link: https://cotap.org/

CARBON EMISSION REMOVAL PROJECTS

Carbon removal is the process of removing carbon dioxide from the atmosphere and locking it away for decades, centuries, or millennia. This could slow, limit, or even reverse climate change-but it is not a substitute for cutting greenhouse gas emissions.

PROMINENT METHODS FOR CARBON REMOVAL:

Afforestation / Reforestation - planting massive new forests.

Soil carbon sequestration - us no-till agriculture and other practices to increase the amount of carbon stored in soils.

Biochar- creating charcoal and burying i t or plowing it into fields.

Bioenergy with CCS or BECCS -capturing and sequestering carbon from biofuels and bioenergy plants.

Enhanced Mineralization - spreading crushed rocks over land to absorb carbon dioxide from the air or exposing them to carbon dioxide-rich fluids.

Direct air capture - building machines that would suck carbon dioxide directly out of the atmosphere and bury it.

Reforestation, or restoring forest ecosystems after they've been damaged by wildfire or cleared for agricultural or commercial uses.

Restocking, or increasing the density of forests where trees have been lost due to disease or disturbances.

Silvestre, or incorporating trees into animal agriculture systems.

Cropland agroforestry, or incorporating trees into row crop agriculture systems.

Urban reforestation, or increasing tree cover in urban areas.

6 WAYS TO REMOVE CARBON POLLUTION FROM THE ATMOSPHERE

1) TREES AND FORESTS

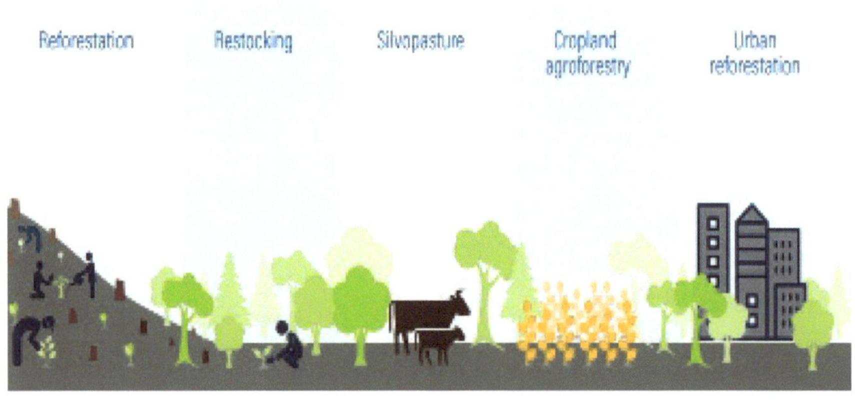

Plants remove carbon dioxide from the air naturally, and trees are especially good at storing CO_2 removed from the atmosphere by photosynthesis. Expanding, and managing tree cover to encourage more carbon uptake can leverage the power of photosynthesis, converting carbon dioxide in the air into carbon stored in wood and soils.

2) FARMS AND SOILS

Soils naturally sequester carbon, but agricultural soils are running a big deficit due to frequent plowing and erosion from farming and grazing, all of which release stored carbon. Because agricultural land is so expansive — encompassing more than 900 million acres in the United States alone, or approximately 40% of the country's land area — even small increases in soil carbon per acre could be impactful. Increasing soil carbon can benefit farmers and   ranchers in addition to removing carbon from the atmosphere.

3) BIOMASS CARBON REMOVAL AND STORAGE

Biomass carbon removal and storage (Bic RS) includes a range of processes that use biomass from plants or algae to remove carbon dioxide from the air and then store it for long periods of time. These methods aim to leverage the carbon storage capacity of plants beyond their natural lifecycles: Whereas trees remove and store carbon only until they die and decompose, biomass carbon removal and storage aims to sequester the CO_2 that plants capture more permanently.

4) DIRECT AIR CAPTURE

Direct air capture is the process of chemically scrubbing carbon dioxide from the ambient air and then sequestering it either underground or in long-lived products like concrete.

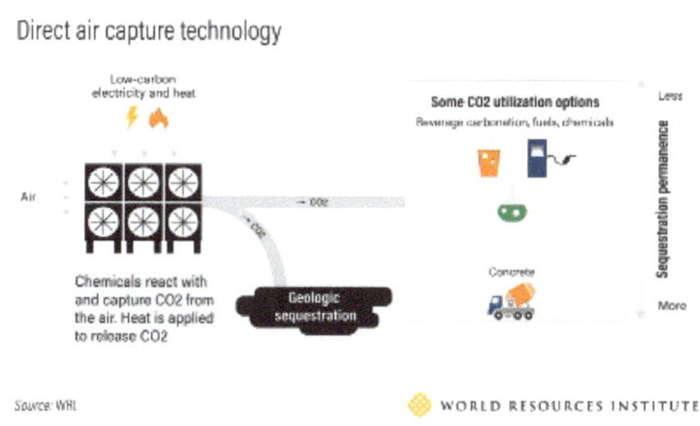

This technology is similar to the carbon capture and storage technology used to reduce emissions from sources like power plants and industrial facilities. The difference is that direct air capture removes excess carbon that's already been emitted into the atmosphere, instead of capturing it at the source.

5) CARBON MINERALIZATION

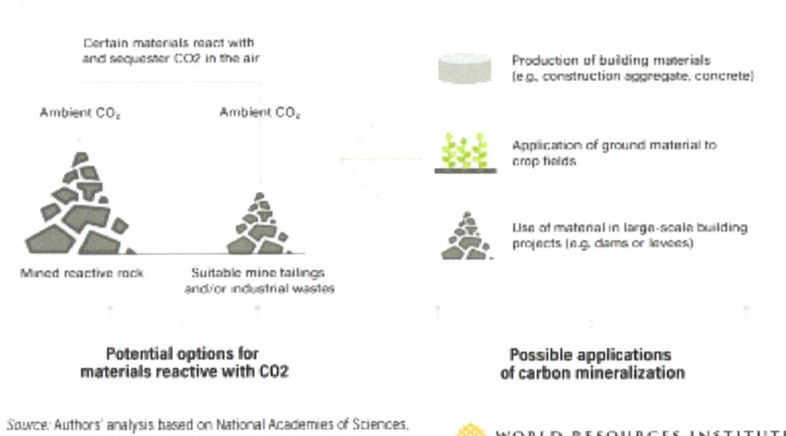

Some minerals naturally react with CO_2, turning carbon dioxide from a gas into a solid and keeping it out of the atmosphere permanently. This process is commonly referred to as "carbon mineralization" or "enhanced weathering," and it naturally happens very slowly, over hundreds or thousands of years.

6) OCEAN-BASED APPROACHES

A number of ocean-based carbon removal approaches have been proposed to leverage the ocean's capacity to sequester carbon and expand the portfolio of options beyond land-based applications. However, nearly all of these strategies are at early stages of development and require more research, and in some cases field testing, to understand whether they are appropriate for investment given potential ecological, social and governance impacts.

The ocean may offer potential carbon removal options, like seaweed cultivation, that could also have ecological benefits. Photo by the National Parks Service.

Reference link: https://www.wri.org/insights/6-ways-remove-carbon-pollution-sky

RENEWABLE ENERGY CERTIFICATES

Renewable Energy Certificates (RECs) are a mechanism for incentivising producers of electricity from renewable energy sources. The relevant regulations have been put   in   place by the Central Electricity Regulatory Commission (CERC).Renewable Energy Certificates (RECs),also known as Green tags, Renewable Energy Credits, Renewable Electricity Certificates, or Tradable Renewable Certificates (TRCs), are tradable, non-tangible energy certificates in the United States that represent proof that 1 megawatt-hour (MWh) of electricity was generated from an eligible renewable energy resource (renewable electricity) and was fed into the shared system of power lines which transport energy. Solar renewable energy certificates (SRECs) are RECs that are specifically generated by solar energy.

WHO ISSUES A REC CERTIFICATE

Central Electricity Regulatory Commission (CERC) has notified Regulation on Renewable Energy Certificate (REC) in fulfillment of its mandate to promote renewable sources of energy and development of market in electricity. The framework of REC is expected to give push to RE capacity addition in the country.

WHO CREATED RENEWABLE ENERGY CERTIFICATES

The Renewable Energy Certificate System (RECS) was a voluntary system for international trade in renewable energy certificates that was created by RECS International to stimulate the international development of renewable energy.

WHAT IS THE LEGAL BASIS OF RECS

RECs are the accepted legal instrument through which renewable energy generation and use claims are substantiated in the U.S. renewable electricity market. RECs aresupported by several different levels of government, regional electricity transmission authorities, nongovernmental organizations (NGOs), and trade associations, as well as in U.S. case law.

WHAT IS REC ARBITRAGE

REC Arbitrage is a green power procurement strategy used by electricity consumers to simultaneously meet two objectives: 1) decrease the cost of their renewable electricity use and 2) substantiate renewable electricity use and carbon footprint reduction claims. The strategy is used by consumers installing self-financedrenewable electricity projects or consumers who purchase renewable electricity directly from a renewable electricity project, such as through a power purchase agreement (PPA).

Source: https://www.kyos.com/wp-content/uploads/2020/10/10-Renewable-energy-certificates-for-mailchimp.jpg
Reference link: https://www.incorp.asia/blogs/recs-vs-carbon-credits-singapore/#:~:text=RECs%20primarily%20target%20Scope%202,across%20Scope%201%20to%203

RENEWABLE ENERGY SOURCES

Renewable energy is energy derived from natural sources that are replenished at a higher rate than they are consumed. Sunlight and wind, for example, are such sources that are constantly being replenished. Renewable energy sources are plentiful and all around us.

HERE ARE A FEW COMMON SOURCES OF RENEWABLE ENERGY:

1.SOLAR ENERGY

Solar energy is the most abundant of all energy resources and can even be harnessed in cloudy weather. The rate at which solar energy is intercepted by the Earth is about 10,000 times greater than the rate at which humankind consumes energy.

2.WIND ENERGY

Wind energy harnesses the kinetic energy of moving air by using large wind turbines located on land (onshore) or in sea- or freshwater (offshore). Wind energy has been used for millennia, but onshore and offshore wind energy technologies have evolved over the last few years to maximize the electricity produced - with taller turbines and larger rotor diameters.

3.GEOTHERMAL ENERGY

Geothermal energy utilizes the accessible thermal energy from the Earth's interior. Heat is extracted from geothermal reservoirs using wells or other means.Reservoirs that are naturally sufficiently hot and permeable are called hydrothermal reservoirs, whereas reservoirs that are sufficiently hot but that are improved with hydraulic stimulation are called enhanced geothermal systems.Once at the surface, fluids of various temperatures can be used to generate electricity. The technology for electricity generation from hydrothermal reservoirs is mature and reliable, and has been operating for more than 100 years.

4.HYDROPOWER

Hydropower reservoirs often have multiple   uses - providing drinking water, water for irrigation, flood and drought control, navigation services, as well as energy supply.Hydropower currently is the largest source of renewable energy in the electricity sector. It relies on generally stable rainfall patterns, and can be negatively impacted by climate-induced droughts or changes to ecosystems which impact rainfall patterns.

5.BIOENERGY

Bioenergy is produced from a variety of organic materials, called biomass, such as wood, charcoal, dung and other manures for heat and power production, and agricultural crops for liquid biofuels. Most biomass is used in rural areas for cooking, lighting and space heating, generally by poorer populations in developing countries.

6.OCEAN ENERGY

Ocean energy derives from technologies that use the kinetic and thermal energy of seawater - waves or currents for instance - to produce electricity or heat.Ocean energy systems are still at an early stage of development, with a number of prototype wave and tidal current devices being explored.

Reference link: https://www.un.org/en/climatechange/what-is-renewable-energy

ZERO CO$_2$ EMISSION VEHICLES

Zero-emission vehicles (ZEVs) are a cleaner and more sustainable alternative to traditional gasoline and diesel-powered vehicles. They produce zero tailpipe emissions, which means they have a significantly lower carbon footprint and can help reduce greenhouse gas emissions and mitigate the effects of climate change. Zero-Emission Vehicles, also known as ZEVs, are vehicles that do not directly emit exhaust gas or other pollutants. ZEV are battery powered and must be plugged in to be recharged (fully electric). Hybrid vehicles use both gas and electricity to operate and have both plug-in and non-plug-in options.

1.BATTERY-ELECTRIC VEHICLES

BEVs use electric motors that draw electricity from on-board rechargeable batteries. They are the most fuel-efficient vehicles available. BEVs produce no tailpipe emissions.

These types of vehicles could be ideal for someone who wants to dramatically reduce fuel costs, greenhouse gas emissions, and vehicle maintenance expenses.

2.PLUG-IN HYBRID ELECTRIC VEHICLES

PHEVs have rechargeable batteries and gas engines, so you have the flexibility of charging at home, or filling up at a conventional gas station. When operating in electric-only mode, PHEVs produce no tailpipe emissions.

This type of vehicle could be ideal for someone who drives long distances and wants to transition to a BEV in the future.

3.HYDROGEN FUEL CELL VEHICLES

FCVs use hydrogen to power the electric motor that propels the vehicle. FCVs emit water vapour and warm air only—they produce no tailpipe emissions.

WHO IS MOVING TO NET ZERO?

The majority of countries have already set targets, or committed to do so, for reaching net zero emissions on timescales compatible with the Paris Agreement temperature goals. Some of those who committed earliest include the UK, Germany, France, Spain, Norway, Denmark, Switzerland, Portugal, New Zealand, Chile, Costa Rica (2050), Sweden (2045), Iceland, Austria(2040) and Finland (2035). The tiny Himalayan Kingdom of Bhutan and the most forested country on earth, Suriname, are already carbon-negative – they absorb more CO_2 than they emit.

FOSSIL FUEL

Fossil fuels are made from decomposing plants and animals. These fuels are found in Earth's crust and contain carbon and hydrogen, which can be burned for energy. Coal, oil, and natural gas are examples of fossil fuels. Coal is a material usually found in sedimentary rock deposits where rock and dead plant and animal matter are piled up in layers. More than 50 percent of a piece of coal's weight must be from fossilized plants. Oil is originally found as a solid material between layers of sedimentary rock, like shale. This material is heated in order to produce the thick oil that can be used to make gasoline. Natural gas is usually found in pockets above oil deposits. It can also be found in sedimentary rock layers that don't contain oil. Natural gas is primarily made up of methane. Fossil fuel, any of a class of hydrocarbon-containing materials of biological origin occurring within Earth's crust that can be used as a source of energy.

Fossil fuels include coal, petroleum, natural gas, oil shales, bitumen's, tar sands, and heavy oils. All contain carbon and were formed as a result of geologic processes acting on the remains of organic matter produced by photosynthesis, a process that began in the Archean Eon (4.0 billion to 2.5 billion years ago). Most carbonaceous material occurring before the Devonian Period (419.2 million to358.9 million years ago) was derived from algae and bacteria, whereas most carbonaceous material occurring during and after that interval was derived from plants. All fossil fuels can be burned in air or with oxygen derived from air to provide heat. This heat may be employed directly, as in the case of home furnaces, or used to produce steam to drive generators that can supply electricity. In still other cases—for example, gas turbines used in jet aircraft—the heat yielded by burning a fossil fuel serves to increase both the pressure and the temperature of the combustion products to furnish motive power.

Since the beginning of the Industrial Revolution in Great Britain in the second half of the 18th century, fossil fuels have been consumed at an ever-increasing rate. Today they supply more than 80 percent of all the energy consumed by the industrially developed countries of the world. Although new deposits continue to be discovered, the reserves of the principal fossil fuels remaining on Earth are limited. The amounts of fossil fuels that can be recovered economically are difficult to estimate, largely because of changing rates of consumption and future value as well as technological developments. In addition, as recoverable supplies of conventional (light-to-medium) oil became depleted, some petroleum-producing companies shifted to extracting heavy oil, as well as liquid petroleum pulled from tar sands and oil shales. See also coal mining; petroleum production.

CONTRIBUTIONS TO GLOBAL WARMING

One of the main by-products of fossil fuel combustion is carbon dioxide (CO_2). The ever-increasing use of fossil fuels in industry, transportation, and construction has added large amounts of CO_2 to Earth's atmosphere. Atmospheric CO_2 concentrations fluctuated between 275 and 290 parts per million by volume (ppmv) of dry air between 1000 CE and the late 18th century but increased to 316 ppmv by 1959 and rose to 421 ppmv in 2023. CO_2 behaves as a greenhouse gas—that is, it absorbs infrared radiation (net heat energy) emitted from Earth's surface and reradiates it back to the surface. Thus, the substantial CO_2 increase in the atmosphere is a major contributing factor to human-induced global warming. Methane (CH4), another potent greenhouse gas, is the chief constituent of natural gas, and CH4 concentrations in Earth's atmosphere rose from 722 parts per billion (ppb) before 1750 to 1,859 ppb by 2018, before rising substantially to 1,919 ppb by 2023.

By the early 21st century, fossil fuels were providing roughly 80 percent of the world's energy. Given the increasing risk posed to Earth's climate by rising concentrations of greenhouse gases in the atmosphere, representatives from nearly 200 countries gathering in Dubai in 2023 at the 28th Conference of the Parties to the United Nations Framework Convention on Climate Change (COP28) agreed to begin to transition the world's economies from fossil fuels to renewable energy. In order to achieve the goal of net-zero carbon emissions by 2050, which would do much to limit average warming worldwide to about 1.5 °C (2.7 °F) above preindustrial levels, the delegates urged countries to accelerate the build-out of solar, wind, and other renewable energy projects, with the objective of tripling renewable energy capacity by 2030.

NON-FOSSIL FUEL ENERGY

Non-fossil fuel energy, also known as renewable energy, is generated from natural resources that can be replenished over time, such as:
1. Solar Energy: Energy generated from the sun's rays, either through photovoltaic (PV) panels or solar thermal systems.
2. Wind Energy: Energy generated from the wind using wind turbines.
3. Hydro Energy: Energy generated from the movement of water in rivers, oceans, or tidal currents, using hydroelectric power plants or tidal power turbines.
4. Geothermal Energy: Energy generated from the heat of the Earth's core, used to produce electricity or provide heating and cooling.
5. Biomass Energy: Energy generated from organic matter such as wood, crops, and waste, through combustion, anaerobic digestion, or gasification.
6. Hydrogen Energy: Energy generated from the reaction of hydrogen with oxygen, potentially produced from renewable energy sources.
7. Tidal Energy: Energy generated from the rise and fall of ocean tides, using tidal barrages or tidal stream generators.
8. Wave Energy: Energy generated from the movement of ocean waves, using wave energy converters.
9. Biofuels: Fuels produced from organic matter such as plants, algae, or agricultural waste, which can be used to power vehicles or generate electricity.
10. Green Gas: A mixture of methane and carbon dioxide, produced from biomass or agricultural waste, which can be used as a substitute for fossil fuels.
Non-fossil fuel energy offers several benefits, including:
1. Reduced greenhouse gas emissions: Decreasing reliance on fossil fuels helps mitigate climate change.
2. Improved air quality: Renewable energy sources produce little to no air pollutants, improving public health.
3. Enhanced energy security: Diversifying energy sources reduces dependence on imported fuels, enhancing energy security.
4. Job creation and economic growth: The renewable energy industry is creating new job opportunities and driving economic growth.
5. Sustainable development: Non-fossil fuel energy supports sustainable development by reducing environmental impacts and promoting resource efficiency. As the world transitions to a low-carbon economy, non-fossil fuel energy will play an increasingly important role in meeting our energy needs while protecting the environment.

Reference link: https://education.nationalgeographic.org/resource/fossil-fuels/ https://www.britannica.com/explore/savingearth/wp-content/uploads/sites/4/2019/03/fossil-fuel-hero.jpg
Reference link:https://education.nationalgeographic.org/resource/non-renewable-energy/ & Meta AI

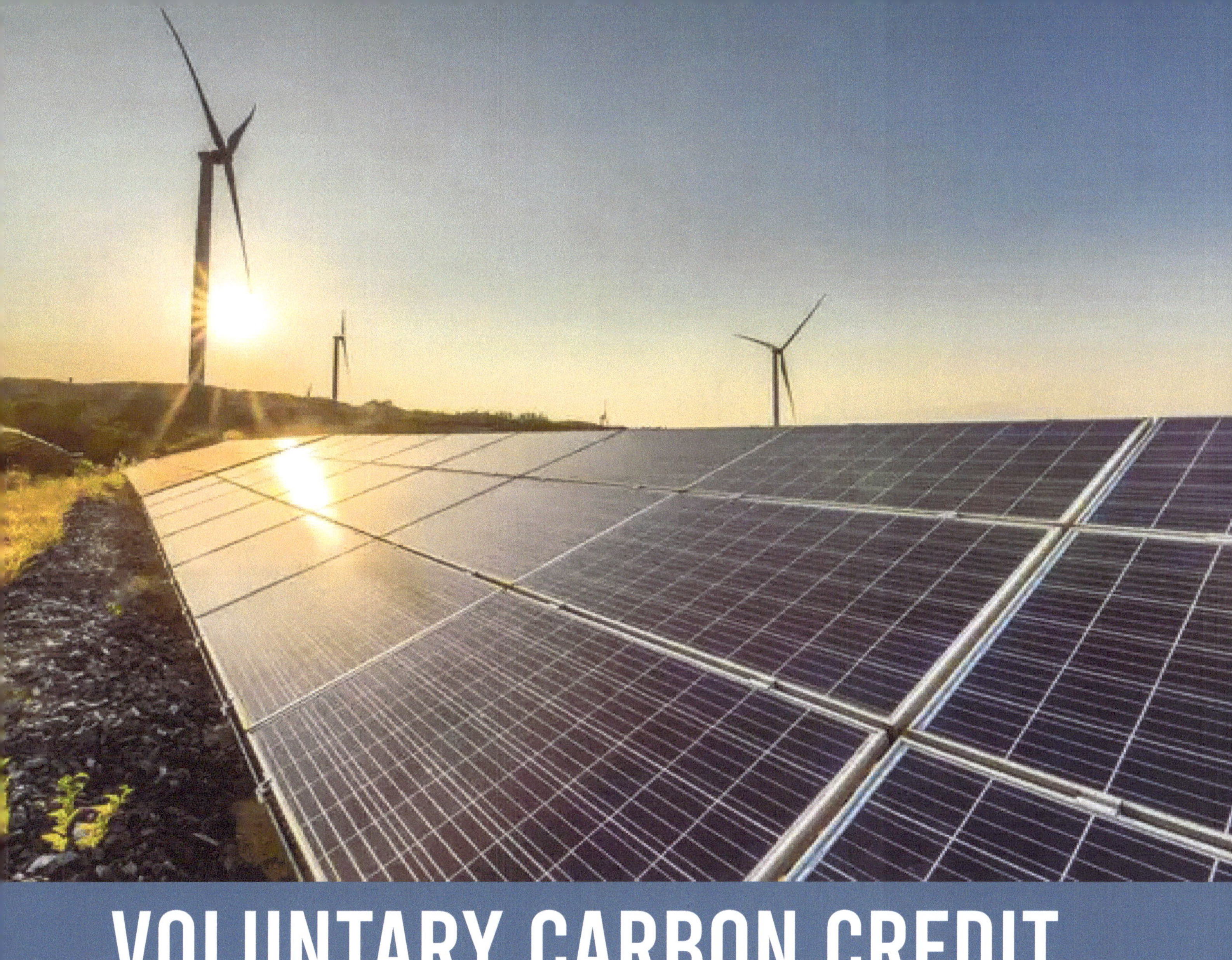

VOLUNTARY CARBON CREDIT

Carbon credits are permits that allow the owner to emit a certain amount of carbon dioxide or other greenhouse gases (GHGs). One credit allows the emission of one ton of carbon dioxide or the equivalent of other greenhouse gases. Carbon credits are also known as carbon allowances. The ultimate goal of the carbon credit system is to reduce the emission of GHGs into the atmosphere.

WHO CAN SELL CARBON CREDITS?

Carbon credits can only be sold or purchased by businesses and governments. Carbon offsets, however, are carbon credits available on the voluntary carbon market. The voluntary carbon market enables entities participating in an emissions reduction project to sell credits that are not regulatory in nature. Anyone can purchase these credits. Carbon credits are sold by governments to businesses, and can be resold on the regulated carbon credit market. Carbon offsets are sold on the voluntary carbon credit market by organizations, projects, or individuals to fund their green projects.

A diverse range of enterprises and Individuals can sell these carbon offsets depending on their ability to participate in a carbon registry or sequestration program. For example, landowners may be able to sell carbon credits if they enroll their land into a project, whether it's reforestation, afforestation, or other carbon removal initiatives, and use the funds to pay for their operations.

Voluntary carbon credits are tradable certificates that represent the reduction or removal of one metric ton of carbon dioxide (or its equivalent) from the atmosphere. These credits are part of the voluntary carbon market, where businesses, organizations, and individuals can purchase them to offset their carbon emissions, even if they are not required to do so by law.

KEY POINTS ABOUT VOLUNTARY CARBON CREDITS:

VOLUNTARY NATURE:

Unlike compliance markets, which are regulated by governments (e.g., the European Union Emissions Trading System), voluntary carbon credits are purchased by entities that want to take proactive steps to reduce their carbon footprint.

PROJECTS:

These credits are typically generated from projects that either reduce emissions (like renewable energy projects) or remove carbon from the atmosphere (such as reforestation or carbon capture initiatives).

VERIFICATION:

To ensure legitimacy, these credits must be verified by third-party organizations that assess the project's effectiveness in reducing or removing carbon emissions.

USES:

Companies often buy these credits to support their sustainability goals, aiming to become carbon neutral or to enhance their environmental credentials.

MARKET DYNAMICS:

The voluntary carbon market is growing as more companies and individuals become aware of climate change and want to contribute to mitigation efforts, even beyond what is legally required.

VOLUNTARY EMISSION REDUCTION

Voluntary Emission Reductions (VER) are emission reductions made voluntarily and not mandated by any regulation or legislation. They usually originate from the will of an organization to take proactive climate action. The voluntary market functions outside of the compliance market. Businesses, organizations, and individuals that wish to offset with no regulatory obligation can use Voluntary Emission Reductions. The carbon credits generated under the VER cannot be used in meeting governmental compliance measures as stated by the Kyoto Protocol. Purpose: Cutting emissions under the voluntary emission scheme will guarantee the quality of your reductions and ensure that they are generated as credible, tradable and recognizable greenhouse gas offsets or credits. DNV will act as your verifier to help ensure that you achieve your reduction objectives.

HOW VER CERTIFICATES ARE ISSUED

VER certificates are based on the principle of compensation. A company can offset greenhouse gases it emits in one place somewhere else. This is made possible through the continuously growing number of climate protection projects that are certified accordingly. For a reforestation project in South America, for example, one VER certificate is issued for each tonne of CO_2 that this forest can compensate. The revenue from the certificates is used by the project investor to fund the climate protection project. All projects and their CO_2-saving volumes are verified according to strict criteria in recognised standards and by neutral auditors. The papers are traded via the international financial markets and trading companies, RWE Supply & Trading being one of them. Buyers, including RWEST's business customers, have exclusive access to the certificates via these channels.. How the VERs are priced depends on various factors, e.g. size of the project, resources and local infrastructure. The more promising, beneficent or influential a project is, the higher the demand for the certificates associated with it, so supply and demand also impact the price. A serial number is issued for every certificate, which is then registered and validated. Furthermore, VER certificates are subject to the principle of additionality, meaning that it has been verified during the certification process that the emissions would not have been saved without the project. The revenues from selling the certificates are thus used for implementing climate protection projects and helping to achieve their targets.

CERTIFIED EMISSION REDUCTION

A Certified Emission Reductions (CER) is a unit of greenhouse gas emission reductions issued pursuant to the Clean Development Mechanism of the Kyoto Protocol, and measured in metric tons of carbon dioxide equivalent. Temporary CERs and Long CERs are special types of CERs issued for forestry projects. They are two ways of accounting for non-permanence in forestry CDM project activities. Temporary CER or TCER is a CER issued for an afforestation or reforestation project activity under the CDM which expires at the end of the commitment period following the one during which it was issued. Long-term CER or ICER is a CER issued for an afforestation or reforestation project activity which expires at the end of its crediting period.

References link: https://plana.earth/glossary/voluntary-emission-reductions-ver#:~:text=Voluntary%20Emission%20Reductions%20(VER)%20are,outside%20of%20the%20compliance%20market. https://www.rwe.com/en/the-group/rwest/rwest-products-and-services/voluntary-emission-reductions-ver-certificates/ Reference link: https://worldrainforests.com/carbon-lexicon/Certified-Emission-Reduction.html

EMISSION TRADING SYSTEM

Emissions trading, also known as 'cap and trade', is a cost-effective way of reducing greenhouse gas emissions. To incentivize firms to reduce their emissions, a government sets a cap on the maximum level of emissions and creates permits, or allowances, for each unit of emissions allowed under the cap. Emitting firms must obtain and surrender a permit for each unit of their emissions. They can obtain permits from the government or through trading with other firms. The government may choose to give the permits away for free or to auction them. Firms that expect not to have enough permits must either cut back on their emissions or buy permits from another firm. For a given permit price, some firms will find it easier, or cheaper, to reduce emissions than others and will sell permits. If there are too many such firms in the market, the price of permits, the total number of which is set in advance by the cap, will decline, inducing some firms to reduce their emissions reduction efforts. Only when the price of permits is just right will the number of permits offered for sale by firms that can reduce emissions at low cost be equal to the number of permits demanded by firms for which emissions reductions are costly. This process of trading ensures there is a unique price for all firms coordinating their activities and drives down emissions to the level allowed under the cap cost-effectively.

WHAT ROLE DOES EMISSIONS TRADING PLAY IN REDUCING CLIMATE CHANGE?

Emissions trading is widely considered a key part of efforts to reduce the manmade greenhouse gas emissions that are causing climate change. The setting of caps is informed by scientific evidence of the emissions cuts needed to limit climate change, including meeting the Paris Agreement target of keeping temperature rise well below 2°C this century. For example, the European Union describes its emissions trading system – the world's largest – as 'a cornerstone' of its climate change policy. It attributes past success in reducing emissions to the system and predicts that in 2020 emissions from the sectors it covers will be 21% lower than in 2005. Research has also shown that the EU emissions trading system has helped to drive innovation in low-carbon technologies such as renewable power sources and energy efficiency, one of the original objectives of the system. Increased use of these technologies also helps to reduce greenhouse gas emissions.

THE INDIAN CARBON MARKET: PATHWAY TOWARDS AN EFFECTIVE MECHANISM

India has pledged to meet its Nationally Determined Contribution (NDC) targets by 2030. Additionally, it aims to attain net-zero emissions by 2070, following the guidelines set forth by the United Nations Framework Convention on Climate Change (UNFCCC). In order to achieve these goals, India is actively exploring decarbonization strategies and employing emission reduction tools and mechanisms for greenhouse gas (GHG) emitting sectors in the country.

This report aims to collate a clear set of learnings from the past and ongoing compliance-based emission trading schemes (ETS) worldwide, including those operating in India. The report includes case studies of European Union ETS and Asian markets like South Korean ETS and China ETS. It also analyses the Surat ETS for PM emissions and India's national Perform Achieve & Trade (PAT) scheme.

The report highlights the lessons from the PAT scheme that needs to be carefully taken into account as the forthcoming Carbon Credit and Trading Scheme (CCTS) is set to transition from the PAT scheme.

WHAT ARE THE BENEFITS OF CARBON MARKETS?

FINANCIAL INCENTIVES:

Carbon markets establish a financial incentive system where entities are allotted emission limits and can trade emission permits. This encourages companies to reduce emissions below their limits and penalizes excess emissions.

COST-EFFECTIVE REDUCTIONS:

Carbon markets prioritize cost-effective emission reductions. Companies that can reduce emissions more easily and at a lower cost are incentivized to do so, leading to overall emission reductions at a lower economic cost.

BUSINESS FLEXIBILITY:

Carbon markets provide businesses with flexibility in choosing how to reduce emissions. They can invest in cleaner technologies, improve energy efficiency, or purchase carbon credits from emission reduction projects elsewhere, allowing for a diverse range of strategies.

CLEAN TECH PROMOTION:

These markets stimulate the development and adoption of cleaner technologies and practices. Companies are motivated to innovate and invest in technologies that reduce emissions to lower their compliance costs in the carbon market.

SUPPORT FOR SUSTAINABILITY:

Carbon markets generate funds for sustainable projects that reduce emissions, such as renewable energy, afforestation, reforestation, and energy efficiency projects. These projects earn carbon credits that can be sold in the market, attracting investments.

CLIMATE GOAL ALIGNMENT:

Carbon markets can be tailored to align with a country's climate goals and international commitments, helping nations meet their emission reduction targets, such as those set in the Paris Agreement, by creating a mechanism for tracking and reducing emissions.

Reference link:
https://www.drishtiias.com/daily-updates/daily-news-editorials/carbon-markets-in-india-a-catalyst-for-green-growth
https://www.cseindia.org/the-indian-carbon-market-pathway-towards-an-effective-mechanism-12328
https://www.lse.ac.uk/granthaminstitute/explainers/how-do-emissions-trading-systems-work/

VOLUNTARY CARBON MARKET

The VCM gives companies, non-profit organizations, governments, and individuals the opportunity to buy and sell carbon offset credits. A carbon offset is an instrument that represents the reduction of one metric tons of carbon dioxide or GHG emissions. To put this in perspective, to capture one ton of CO_2 emissions you would have to grow approximately 50 trees for one-year [1].

Companies that are unable to reach their greenhouse gas (GHG) emission targets can purchase carbon offset credits by investing in environmental projects that can avoid, reduce, or remove carbon emissions.

VOLUNTARY VS. COMPLIANCE MARKET

	VOLUNTARY MARKET	COMPLIANCE MARKET
Exchanged Commodity Carbon offsets.	Facilitated by the project-based system	Allowances. Facilitated by the cap-and-trade system.
How is the market regulated?	Functions outside of the compliance market.	National, regional or international carbon reduction regimes E.g. Kyoto Protocol, California Carbon Market
What is the price?	Voluntary credits tend to be cheaper because they cannot be used in compliance markets.2 Several factors impact the price such as project type, project size, location, co-benefits, and vintage.	Compliance credits tend to be more expensive because they are driven by regulatory obligations.
Who can purchase credits?	Businesses, governments, NGOs, and individuals	Companies and governments have adopted emission limits established by the United Nations Convention on Climate change
Where do credits trade?	Currently no centralized voluntary carbon credit market. Project developers can sell credits directly to buyers, through a broker or an exchange, or sell to a retailer who then resells to a buyer.	Companies that surpass their emission targets can sell their surplus credits to those looking to offset emissions. Credits can be sold under the Kyoto Protocols emissions trading scheme.

Reference link: https://carboncredits.com/what-is-the-voluntary-carbon-market/

VOLUNTARY CARBON MARKET

 "Net zero" emissions by 2050 is a goal that needs to be achieved when it comes to fighting climate change and which requires change by the entire economy. To reach the "net-zero goals", companies do not only need to decarbonizes but also compensate and increasingly neutralize their emissions. Robust, trustful, and secure voluntary carbon markets are therefore needed to allow offsetting emissions by purchasing carbon credits.Voluntary carbon markets allow carbon emitters to offset their emissions by purchasing carbon credits emitted by projects targeted at removing or reducing greenhouse gas from the atmosphere.

CARBON MARKETS CAN BE BROADLY DIVIDED INTO TWO SEGMENTS:

COMPLIANCE MARKETS:

Driven by binding emission reduction targets (e.g. ETS) or other types of regulation (e.g. tax)

VOLUNTARY MARKETS:

Corporate or individuals who wish to offset their emissions ("voluntary offsetting")Historically, voluntary demand has led to a proliferation of privately governed certification standards. Those standard organizations issue carbon credits for certain project activities (project-based). As a result, different types of carbon credits with different qualities exist. Carbon credits can be broadly grouped into two main categories: reduction credits and removal credits.

Source: https://terrapass.com/blog/voluntary-carbon-market-how-participate/ Reference link: https://www.eex.com/en/markets/environmental-markets/voluntary-carbon-markets

NET ZERO – NOT ZERO EMISSION

Net-zero emissions, or "net zero," will be achieved when all emissions released by human activities are counterbalanced by removing carbon from the atmosphere in a process known as carbon removal Achieving net zero will require a two-part approach: First and foremost, human-caused emissions (such as those from fossil-fuelled vehicles and factories) should be reduced as close to zero as possible. Any remaining emissions should then be balanced with an equivalent amount of carbon removal, which can happen through natural approaches like restoring forests or through technologies like direct air capture and storage (DACS), which scrubs carbon directly from the atmosphere.

WHEN DOES THE WORLD NEED TO REACH NET-ZERO EMISSIONS?

Under the Paris Agreement, countries agreed to limit warming to well below 2 degrees C (3.6 degrees F), ideally to 1.5 degrees C (2.7 degrees F). Global climate impacts that are already unfolding under the current 1.1 degrees C (1.98 degrees F) of warming — from melting ice to devastating heat waves and more intense storms — show the urgency of minimizing temperature increase. The latest science suggests that limiting warming to 1.5 degrees C depends on CO_2 emissions reaching net zero between 2050 and 2060. Reaching net zero earlier in that range (closer to 2050) avoids a risk of temporarily "overshooting," or exceeding 1.5 degrees C. Reaching net zero later (nearer to 2060) almost guarantees surpassing 1.5 degrees C for some time before global temperature can be reduced back to safer limits through carbon removal. Critically, the sooner emissions peak, and the lower they are at that point, the more realistic achieving net zero becomes. This would also create less reliance on carbon removal in the second half of the century. This does not suggest that all countries need to reach net-zero emissions at the same time. However, the chances of limiting warming to 1.5 degrees C depend significantly on how soon the highest emitters reach net zero. Equity-related considerations — including responsibility for past emissions, equality in per-capita emissions and capacity to act — also suggest earlier dates for wealthier, higher-emitting countries. Importantly, the time frame for reaching net-zero emissions is different for CO_2 alone versus for CO_2 plus other greenhouse gases like methane, nitrous oxide and fluorinated gases. For non-CO_2 emissions, the net zero date is later, in part because models assume that some of these emissions — such as methane from agricultural sources — are more difficult to phase out. However, these potent but short-lived gases will drive temperatures higher in the near term, potentially pushing temperature change past the 1.5 degrees C threshold much earlier. Because of this, it's important for countries to specify whether their net-zero targets cover CO_2 only or all GHGs. A comprehensive net-zero emissions target would include all GHGs, ensuring that non-CO_2 gases are also reduced with urgency.

NET ZERO TRANSITION

The just transition is a crucial enabler to implementing the net zero transition: involving all affected parties and responding to injustices serves to ensure political acceptability for climate action, mitigate the risk of 'just transition litigation', and ultimately avoid delays in achieving net zero globally. To achieve its varied aims, a just transition is considered to require fundamental restructuringof the socioeconomic systems that have created these inequalities and the climate crisis.

WHO DOES THE JUST TRANSITION PRIORITISE AND HOW?

The just transition seeks to centre the interests of those that are most affected by the low-carbon transition, including workers, vulnerable communities, suppliers of goods and services, specifically small and medium-sized enterprises (SMEs), and consumers. This approach strongly advocates the inclusion of these stakeholders in shaping the net zero transition so that no one is 'left behind'.

Successful transition planning would involve these stakeholders, for example through ongoing participation processes and channels, including social forums, consultations and citizens' assemblies.

SUSTAINABILITY

Sustainability refers to the ability to maintain or support a process continuously over time. In business and policy contexts, sustainability seeks to prevent the depletion of natural or physical resources, so that they will remain available for the long term.Sustainability is ability to maintain or support a process over time.Sustainability is often broken into three core concepts: economic, environmental, and social.Many businesses and governments have committed to sustainable goals, such as reducing their environmental footprints and conserving resources. Some investors are actively embracing sustainability investments, known as "green investments". Skeptics have accused some companies of "greenwashing" the practice of misleading the public to make a business seem more environmentally friendly than it is. Sustainability is a social goal for people to co-exist on Earth over a long period of time. Definitions of this term are disputed and have varied with literature, context, and time. Sustainability usually has three dimensions (or pillars): environmental, economic, and social. Many definitions emphasize the environmental dimension. This can include addressing key environmental problems, including climate change and biodiversity loss. The idea of sustainability can guide decisions at the global, national, organizational, and individual levels. A related concept is that of sustainable development, and the terms are often used to mean the same thing. UNESCO distinguishes the two like this: "Sustainability is often thought of as a long-term goal (i.e. a more sustainable world), while sustainable development refers to the many processes and pathways to achieve it.

Source: https://tontoton.com/5-benefits-of-a-circular-economy/ Reference link: https://www.investopedia.com/terms/s/sustainability.asp#:~:text=Sustainability%20is%20ability%20to%20 maintain,environmental%20footprints%20and%20conserving%20resources. https://en.m.wikipedia.org/wiki/Sustainability

PETERSBERG CLIMATE DIALOGUE

The international conference Petersberg Climate Dialogue (PCD) (in German: Petersberger Klimadialog) is a series of negotiations to prepare for the yearly UN Climate Change Conferences in spring or summer between the COP conferences. The appointed next COP-president, with delegation, is usually the co-host of PCD. The first meeting of the PCD was initiated after the nearly completely failed negotiations at the 2009 United Nations Climate Change Conference in Copenhagen (COP15), by German Chancellor Angela Merkel to improve communication with leaders and between the environmental ministers. It was held 2–4 April 2010 in Hotel Petersberg, on the hill named "Petersberg" near the German city of Bonn, the seat of UNFCCC.

In subsequent years the conference took place in Berlin. Every year, Merkel and her successor Olaf Scholz have given a speech and taken part in the discussions. Since 2022, the conference has been hosted by Minister for Foreign Affairs Annalena Baerbock.

The Petersberg Climate Dialogue (25 to 26 April 2024) is an important milestone for the climate negotiations at COP29 in Baku. Representatives from more than 40 countries will come to the Federal Foreign Office for the conference. The road to Baku leads via the Petersberg Climate Dialogue, where the focus is on preparing COP29 in Azerbaijan's capital. Hosted by Foreign Minister Annalena Baerbock and COP President-Designate Mukhtar Babayev, it is an important milestone in the international

Reference link: https://en.wikipedia.org/wiki/Petersberg_Climate_Dialogue https://www.auswaertiges-amt.de/en/aussenpolitik/themen/KlimaEnergie/petersberg-climate-dilogue/2653910 Source:https://www.auswaertiges-amt.de/image/2654770/123x55/1230/550/3df9d1947d086655b45c90a8d4ada440/pA/240425-pcd-bild1.jpg, https://www.auswaertiges-amt.de/image/2654770/123x55/1230/550/3df9d1947d086655b45c90a8d4ada440/pA/240425-pcd-bild1.jpg climate calendar. At the conference, discussions will be held at political level about the key challenges of international climate policy. The global energy transition will only be successful with sufficient private investments in environmentally friendly energies and technologies. Financing the global climate transition will therefore also be a focus of the discussions this year. This year's Petersberg Climate Dialogue will once again feature high-level discussions. In addition to Federal Chancellor Olaf Scholz and Azerbaijani President Ilham Aliyev as the host of COP29, Economics Minister Robert Habeck and Development Minister Svenja Schulze (Federal Ministry for Economic Cooperation and Development) will also speak to representatives from around 40 countries. At the Petersberg Climate Dialogue, decision-makers from industrialised, emerging and developing countries will also go to the drawing board together to lay the groundwork for the upcoming decisions ●

DECARBONISATION

'Decarbonisation' tends to refer to the process of reducing 'carbon intensity', lowering the amount of greenhouse gas emissions produced by the burning of fossil fuels. Generally, this involves dcreasing CO_2 output per unit of electricity generated. It is reducing the amount of carbon dioxide occurring as a result of transport and power generation is essential to meet global temperature standards set by the Paris Agreement and UK government. Decarbonisation is the reduction of carbon dioxide emissions through the use of low carbon power sources, achieving a lower output of greenhouse gasses into the atmosphere ●

How Does Decarbonisation Work?

Decarbonisation involves increasing the prominence of low-carbon power generation, and a corresponding reduction in the use of fossil fuels. This involves in particular a use of renewable energy sources like wind power, solar power, and biomass. The use of carbon power can also be reduced through large-scale use of electric vehicles alongside 'cleaner' technologies. Decreasing carbon intensity in the power and transport sectors will allow for net zero emission targets to be met sooner and in line with government standards ●

Why is Decarbonisation Important?

The UK government committed to achieving net zero greenhouse gas emissions by 2050. After Parliament's declaration of a climate emergency, the Committee on Climate Change recommended that achieving this net zero was not only feasible but also necessary and cost-effective. Rapid decarbonisation is becoming more necessary as the transport sector becomes electrified, increasing the demand for electric power. Greater energy efficiency is therefore becoming a priority to meet emission targets and improve air quality and global temperature ●

Source: https://www.twi-global.com/CachedImage.axd?ImageName=What-is-Decarbonisation.jpg&ImageWidth=784&ImageHeight=379&ImageVersionID=98788&ImageMoffied=20200624101608
Reference link:https://www.twi-global.com/technical-knowledge/faqs/what-is-decarbonisation#:~:text=Decarbonisation%20is%20the%20reduction%20of,greenhouse%20gasses%20into%20the%20atmosphere

CARBON DIOXIDE EQUIVALENTS

Carbon dioxide equivalents (CO$_2$e) are a measure of the effect of different greenhouse gases (GHGs) on the climate. By converting different emissions to the equivalent amount of carbon dioxide (CO$_2$), their impacts can be compared. CO2equivalent (CO2e) is a unit of measurement that is used to standardise the climate effects of various greenhouse gases●

How are CO$_2$e measured?

CO$_2$e are quantified by multiplying the amount of the relevant GHG by its respective GWP. The resulting number shows the amount of carbon dioxide that would need to be emitted to have the same impact on the atmosphere. CO$_2$e are measured by mass. Depending on the application, they might be expressed in kilograms or tonnes●

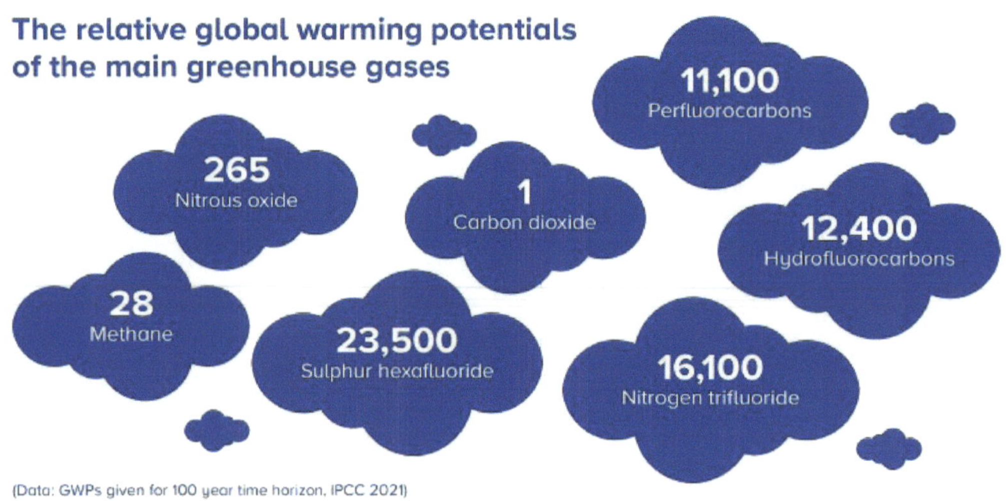

What are CO$_2$e used for?

CO$_2$e are used in a range of contexts where comparison between GHGs is useful or necessary. When quantifying a carbon footprint, it's common to express the total amount of emissions as CO$_2$e, to reflect that not only carbon dioxide is being emitted. For the same reason, emission factors are generally expressed as CO$_2$e. Another standard use for CO$_2$e is carbon credits, which are issued to quantify the impact of climate projects. Carbon credits traded on the voluntary carbon market are called Verified Emission Reductions (VERs). One VER represents ownership of one tonne of carbon dioxide or its equivalent in other greenhouse gases that can be traded, sold, or retired. In order to make the effects of different greenhouse gases comparable, the Inter-governmental Panel on Climate Change (IPCC) of the United Nations has defined the so-called "Global Warming Potential". This index expresses the warming effect of a certain amount of a greenhouse gas over a set period of time (usually 100 years) in comparison to CO$_2$ For example, methane's effect on the climate is 28 times more severe than CO$_2$, but it doesn't stay in the atmosphere as long. The environmental impact of nitrous oxide also exceeds that of CO$_2$ by almost 300 times. The anthropogenic source of these greenhouse gases can be found in agriculture through the use of nitrogen fertilizers and livestock farming. In this way, greenhouse gases can be calculated as CO$_2$ equivalents. CO$_2$ equivalents are abbreviated with "CO$_2$e"●

Source: https://www.climatepartner.com/sites/default/files/content/images/glossary/greenhouse-gases-global-warming-potentials.jpg
Reference link: https://www.climatepartner.com/en/knowledge/glossary/carbon-dioxide-equivalent

POLLUTER PAYS PRINCIPLE (PPP)

The 'polluter pays' principle is the commonly accepted practice that those who produce pollution should bear the costs of managing it to prevent damage to human health or the environment. For instance, a factory that produces a potentially poisonous substance as a by-product of its activities is usually held responsible for its safe disposal. The polluter pays principle is part of a set of broader principles to guide sustainable development worldwide (formally known as the 1992 Rio Declaration). The principle underpins most of The regulation of pollution affecting land, water and air. Pollution is defined in UK law as contamination of the land, water or air by harmful or potentially harmful substances.

WHY SHOULD THE PRINCIPLE BE APPLIED TO GREENHOUSE GAS EMISSIONS?

Greenhouse gas emissions are considered a form of pollution because they cause potential harm and damage through impacts on the climate, and also contribute to outdoor or ambient air pollution, which the World Health Organisation estimates is associated with 4.2 million deaths annually. But because society has been slow to recognise the link between how human activities have increased the rates of greenhouse gases emissions that can cause the climate to change, emitters are generally not held responsible for controlling this form of pollution. When the pollution cost from the release of greenhouse gases is not imposed on emitters, these costs are thus 'externalised' to society, representing what economists describe as a 'market failure'. Society bears these costs as greenhouse gases are emitted into the atmosphere, which is described a 'global commons' as everyone shares and has the right to 'use' it●

Reference link: https://www.lse.ac.uk/granthaminstitute/explainers/what-is-the-polluter-pays-principle/
Source: https://www.lse.ac.uk/granthaminstitute/wp-content/webp-express/webp-images/doc-root/wp-content/uploads/2014/05/pollution.jpg.webp

CLIMATE-RELATED
EXTREME WEATHER EVENTS

Extreme events are occurrences of unusually severe weather or climate conditions that can cause devastating impacts on communities and agricultural and natural ecosystems. Weather-related extreme events are often short-lived and include heat waves, freezes, heavy downpours, tornadoes, tropical cyclones and floods. Climate-related extreme events either persist longer than weather events or emerge from the accumulation of weather or climate events that persist over a longer period of time. Examples include drought resulting from long periods of below-normal precipitation or wildfire outbreaks when a prolonged dry, warm period follows an abnormally wet and productive growing season.Scientists usually define an extreme event using either of two approaches. The first approach examines the probability or chance of an event of a given magnitude occurring within a certain reference period (e.g., 1961 - 1990); here an extreme event has a low probability of occurring at a given location (< 10%) and is typically of a high intensity. This type of probabilistic approach is applied in extreme event attribution to determine whether global warming is driving changes in the frequency and intensity of extreme events ●

Source: https://wpcdn.zenger.news/wp-content/uploads/2021/08/10075512/feat_05362192-5761-41e8-8f26-1d6b4f445f63.jpg
Reference link: https://www.climatehubs.usda.gov/content/extreme-weather

HEAT WAVES

Heat waves are exacerbated when combined with other extreme events. Heat waves can enhance buildup of ozone and other pollutants, leading to combinations of heat and air pollution dangerous for human health. A particularly striking example is the combination of high temperatures, wildfire, and smoke that results in high levels of air pollution across broad geographic areas. In recent summers, the western United States has been afflicted with multiple episodes; these are projected to become increasingly common in the future because of climate change ●

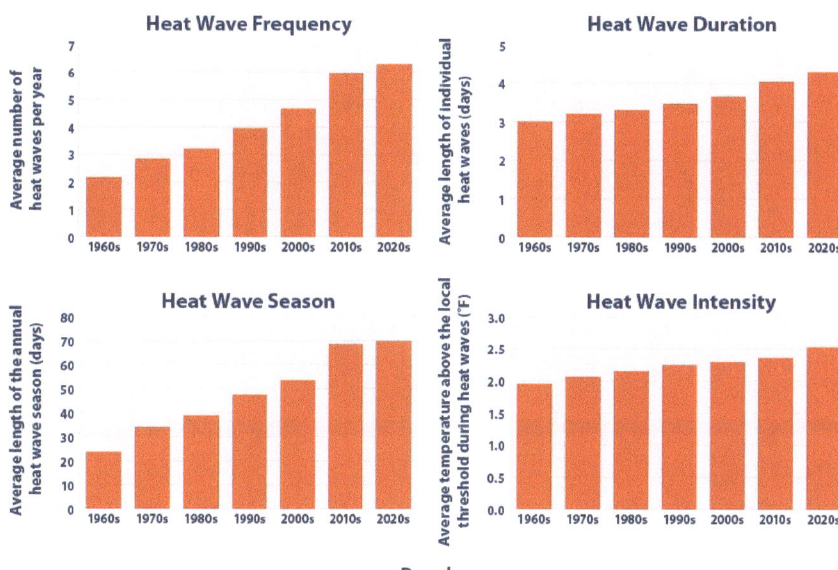

Heat waves are occurring more often than they used to in major cities across the United States. Their frequency has increased steadily, from an average of two heat waves per year during the 1960s to six per year during the 2010s and 2020s.

In recent years, the average heat wave in major U.S. urban areas has been about four days long. This is about a day longer than the average heat wave in the 1960s.

The average heat wave season across the 50 cities in this indicator is about 46 days longer now than it was in the 1960s. Timing can matter, as heat waves that occur earlier in the spring or later in the fall can catch people off-guard and increase exposure to the health risks associated with heat waves.

Heat waves have become more intense over time. During the 1960s, the average heat wave across the 50 cities in Figures 1 and 2 was 2.0°F above the local 85th percentile threshold. During the 2020s, the average heat wave has been 2.5°F above the local threshold.

Source: https://www.epa.gov/system/files/styles/large/private/images/2024-06/heat-waves_figure1_2024.png?itok=ePr_rxK9
Reference link: https://www.epa.gov/climate-indicators/climate-change-indicators-heat-waves#:~:text=Unusually%20hot%20days%20and%20heat,become%20more%20frequent%20and%20intense.

HURRICANES

A hurricane is a large, swirling tropical storm that forms in the open ocean and moves towards land at speeds of over 72 miles per hour. Hurricanes form over the Northeast Pacific or the North Atlantic Ocean. They are known as cyclones in the Indian Ocean and South Pacific, and typhoons when they form in the Northwest Pacific ●

How Are Hurricanes Formed?

Hurricanes begin forming as tropical disturbances in areas of the ocean that are warmer with temperatures of at least 80 °F. These disturbances are because of low pressure caused by the warm seas. If the ocean temperature continues rising, a storm forms. The speed of the storm may vary, but when it reaches 38 miles, it becomes a tropical depression. The tropical depression gradually develops into a tropical storm and acquires a name when it reaches wind speeds of 39 miles. When the storm sustains wind speeds of 74 miles/h, it is declared a hurricane. Hurricanes produce a staggering amount of energy by absorbing the moist and warm ocean air and releasing it through condensation during thunderstorms. Hurricanes revolve around a calm low-pressure center (eye) that can be 20- to 30-mile wide. The spinning part of the hurricane is known as the "eyewall," and it is responsible for the strong winds and rain.

Damaging Effects of Hurricanes

When a hurricane reaches land, it causes damages that can be catastrophic and results in a storm surge that can rise to 20 feet and extend for about 100 miles. In fact, storm surges result in 90% of the deaths caused by hurricanes. The winds not only cause massive damage but can also spawn tornadoes. Excessive rainfall cause floods that may occur miles inland. The best way to defend against a hurricane is to get out of its way; most of the hurricanes are forecasted and warnings issued to the communities that may be affected 24 hours prior to the hurricane. condensation during thunderstorms. Hurricanes revolve around a calm low-pressure center (eye) that can be 20- to 30-mile wide. The spinning part of the hurricane is known as the "eyewall," and it is responsible for the strong winds and rain ●

Reference link: https://www.worldatlas.com/articles/what-is-a-hurricane.html
Source: https://www.worldatlas.com/r/w960-q80/upload/fe/e3/eb/shutter-stock-718981030.jpg

STROMS

Astorm is any disturbed state of the natural environment or the atmosphere of an astronomical body. It may be marked by significant disruptions to normal conditions such as strong wind, tornadoes, hail, thunder and lightning (a thunderstorm), heavy precipitation (snowstorm, rainstorm), heavy freezing rain (ice storm), strong winds (tropical cyclone, windstorm), wind transporting some substance through the atmosphere such as in a dust storm, among other forms of severe weather. Storms have the potential to harm lives and property via storm surge, heavy rain or snow causing flooding or road impassibility, lightning, wildfires, and vertical and horizontal wind shear. Systems with significant rainfall and duration help alleviate drought in places they move through. Heavy snowfall can allow special recreational activities to take place which would not be possible otherwise, such as skiing and snowmobiling.

Storms are created when a center of low pressure develops with the system of high pressure surrounding it. This combination of opposing forces can create winds and result in the formation of storm clouds such as cumulonimbus. Small localized areas of low pressure can form from hot air rising off hot ground, resulting in smaller disturbances such as dust devils and whirlwinds ●

Source: https://outforia.com/wp-content/uploads/2021/06/Types-of-storms-tornadoes-0621.jpg
Reference link: https://en.wikipedia.org/wiki/Storm

HEAVY RAINFALL

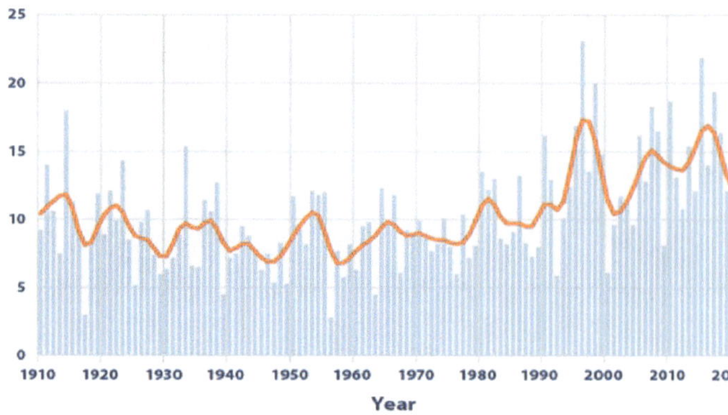

"Heavy precipitation" refers to instances during which the amount of rain or snow experienced in a location substantially exceeds what is normal. What constitutes a period of heavy precipitation varies according to location and season.

Climate change can affect the intensity and frequency of precipitation. Warmer oceans increase the amount of water that evaporates into the air. When more moisture-laden air moves over land or converges into a storm system, it can produce more intense precipitation—for example, heavier rain and snow storms. The potential impacts of heavy precipitation include crop damage, soil erosion, and an increase in flood risk due to heavy rains (see the River Flooding indicator)—which in turn can lead to injuries, drownings, respiratory health impacts from exposure to mold, and other flooding-related effects on health. In addition, runoff from precipitation can impair water quality as pollutants deposited on land wash into water bodies.

Heavy precipitation does not necessarily mean the total amount of precipitation at a location has increased—just that precipitation is occurring in more intense events. However, changes in the intensity of precipitation, when combined with changes in the interval between precipitation events, can also lead to changes in overall precipitation totals ●

In recent years, a larger percentage of precipitation has come in the form of intense single-day events. Nine of the top 10 years for extreme one-day precipitation events have occurred since 1995.

The prevalence of extreme single-day precipitation events remained fairly steady between 1910 and the 1980s, but has risen substantially since then. Over the entire period from 1910 to 2023, the portion of the country experiencing extreme single-day precipitation events increased at a rate of about half a percentage point per decade.

The percentage of land area experiencing much greater than normal yearly precipitation totals increased between 1895 and 2023. There has been much year-to-year variability, however.

In some years there were no abnormally wet areas, while a few others had abnormally high precipitation totals over 10 percent or more of the contiguous 48 states' land area. For example, 1941 was extremely wet in the West, while 1983 was very wet nationwide.

Reference link: https://www.epa.gov/climate-indicators/climate-change-indicators-heavy-precipitation#:~:text=Climate%20
change%20can%20affect%20the,heavier%20rain%20and%20snow%20storms.
Source: https://www.rmets.org/sites/default/files/inline-images/Luo%20Xing%20-%20Thunder%20in%20Chongqing.jpg

DROUGHT

Drought is a prolonged dry period in the natural climate cycle that can occur anywhere in the world. It is a slow-onset disaster characterized by the lack of precipitation, resulting in a water shortage. Drought can have a serious impact on health, agriculture, economies, energy and the environment.

An estimated 55 million people globally are affected by droughts every year, and they are the most serious hazard to livestock and crops in nearly every part of the world. Drought threatens people's livelihoods, increases the risk of disease and death, and fuels mass migration. Water scarcity impacts 40% of the world's population, and as many as 700 million people are at-risk of being displaced as a result of drought by 2030.

Rising temperatures caused by climate change are making already dry regions drier and wet regions wetter. In dry regions, this means that when temperatures rise, water evaporates more quickly, and thus increases the risk of drought or prolongs periods of drought. Between 80-90% of all documented disasters from natural hazards during the past 10 years have resulted from floods, droughts, tropical cyclones, heat waves and severe storms.

IMPACT OF DROUGHT

When drought causes water and food shortages there can be many impacts on the health of the affected population, which may increase the risk of disease and death. Drought may have acute and chronic health effects, including:

● Malnutrition due to the decreased availability of food, including micronutrient deficiency, such as iron-deficiency anaemia;

● Increased risk of infectious diseases, such as cholera, diarrhoea, and pneumonia, due to acute malnutrition, lack of water and sanitation, and displacement;

● Psycho-social stress and mental health disorders;

● Disruption of local health services due to a lack of water supplies, loss of buying power, migration and/or health workers being forced to leave local areas.

Severe drought can also affect air quality by making wildfires and dust storms more likely, increasing health risk in people already impacted by lung diseases, like asthma or chronic obstructive pulmonary disease (COPD), or with heart disease●

Source: https://www.google.com/url?sa=i&url=https%3A%2F%2Fwww.orissapost.com%2Fprolonged-droughts-like-ly-spelled-end-for-indus-magacities-study%2F&psig=AOvVaw3Su7twNDnfd-g9rad4QwGO&ust=1726849685818000&source=imag-es&cd=vfe&opi=89978449&ved=0CBQQjRxqFwoTCPD43N_D0IgDFQAAAAAdAAAAABAJ
Reference link: https://www.who.int/health-topics/drought#tab=tab_2

FLOOD

Flooding is an overflowing of water onto land that is normally dry. Floods can happen during heavy rains, when ocean waves come on shore, when snow melts quickly, or when dams or levees break. Damaging flooding may happen with only a few inches of water, or it may cover a house to the rooftop. Floods can occur within minutes or over a long period, and may last days, weeks, or longer. Floods are the most common and widespread of all weather-related natural disasters.

EFFECTS OF FLOODS

When floodwaters recede, affected areas are often blanketed in silt and mud. This sediment can be full of nutrients, benefiting farmers and agribusinesses in the area. Famously fertile flood plains like the Mississippi River valley in the American Midwest, the Nile River valley in Egypt, and the Fertile Crescent in the Middle East have supported agriculture for thousands of years. Yearly flooding has left millions of tons of nutrient-rich soil behind.

However, floods have enormous destructive power. When a river overflows its banks or the sea moves inland, many structures are unable to withstand the force of the water. Bridges, houses, trees, and cars can be picked up and carried off. Floods erode soil, taking it from under a building's foundation, causing the building to crack and tumble. Severe flooding in Bangladesh in July 2007 led to more than a million homes being damaged or destroyed.

Floods can cause even more damage when their waters recede. The water and landscape can be contaminated with hazardous materials, such as sharp debris, pesticides, fuel, and untreated sewage. Potentially dangerous Mold can quickly overwhelm water-soaked structures. As flood water spreads, it carries disease. Flood victims can be left for weeks without clean water for drinking or hygiene. This can lead to outbreaks of deadly diseases like typhoid, malaria, hepatitis A, and cholera. This happened in 2000, as hundreds of people in Mozambique fled to refugee camps after the Limpopo River flooded their homes. They soon fell ill and died from cholera, which is spread by unsanitary conditions, and malaria, spread by mosquitos.

NATURAL CAUSES OF FLOOD

● Floods occur naturally. They are part of the water cycle, and the environment is adapted to flooding. Wetlands along river banks, lakes, and estuaries absorb flood waters. Wetland vegetation, such as trees, grasses, and sedges, slow the speed of flood waters and more evenly distribute their energy. According to the U.S. Environmental Protection Agency (EPA), the wetlands along the Mississippi River once stored at least 60 days of flood water. (Today, Mississippi wetlands store only 12 days of flood water. Most wetlands have been filled or drained.)

● Floods can also devastate an environment. The most vulnerable regions are those that experience frequent floods and those that have not flooded for many years. In the first case, the environment does not have time to recover between floods In August 2010, Pakistan experienced some of the worst floods of the century. The annual monsoon, on which Pakistani farmers and consumers rely, was unusually strong. Tons of water drenched the nation. The Indus River burst its banks. Because the river flows almost directly through the narrow country, almost all of Pakistan was affected by flooding.

● Millions of Pakistanis lost their homes, and almost 2,000 died in the floods. The province of Punjab, the country's agricultural center, was particularly devastated. Rice, wheat, and corn crops were destroyed. The impact of the floods continued long after the monsoon dwindled and the Indus subsided. Pakistanis experienced food shortages, power outages, and loss of infrastructure. Outbreaks of cholera and malaria developed near resettlement camps. Experts estimated that the rebuilding effort would cost up to $15 billion.

● Sometimes, floods are triggered by other natural disasters, such as earthquakes and tsunamis. In January 2011, a major earthquake struck off the coast of Miyagi Prefecture, Japan. The quake triggered a massive tsunami, its crest reaching as high as 40 meters (131 feet). The tsunami crashed more than 10 kilometers (six miles) inland, flooding homes, businesses, schools, parks, hospitals, and the Fukushima Dai-Ichi Nuclear Power Plant. A dam holding a reservoir burst, triggering another flood that destroyed homes.

● Rain that accompanies hurricanes and cyclones can quickly flood coastal areas. The rise in sea level that occurs during these storms is called a storm surge. A storm surge is a type of coastal flood. They can be devastating. The storm surge that accompanied the 1970 Bhola cyclone flooded the low-lying islands of the Ganges Delta in India and Bangladesh. More than 500,000 people were killed, and twice that number were left homeless.

● The strong winds associated with hurricanes and cyclones can also whip up and move huge amounts of water, forcing a storm surge far inland. In 2005, Hurricane Katrina brought huge amounts of wind and rain to the Gulf Coast of the United States. The city of New Orleans, Louisiana, was particularly hard-hit. The storm surge from Hurricane Katrina caused some of the city's levees to break. Levees protect New Orleans from the Mississippi River. The river rushed in and flooded entire neighbourhoods. Hundreds of people drowned, and the storm did more than $100 billion in damage ●

Source: https://ichef.bbci.co.uk/ace/standard/976/cpsprodpb/D4C0/production/_131946445_gettyimages-1826219275.jpg
Reference link: https://www.nssl.noaa.gov/education/svrwx101/floods/

MONSOON

WHAT IS A MONSOON?

A monsoon is a shift in winds that often causes a very rainy season or a very dry season. Although monsoons are usually associated with parts of Asia, they can happen in many tropical and subtropical regions – including several locations in the United States. Although many of the most well-known monsoons are in Asia, the U.S. Southwest also regularly experiences monsoon season. This photo shows a storm during monsoon season in Petrified Forest National Park in Arizona. Credit: National Park Service/Stewart Holmes When people think of a monsoon, they often think of heavy rains that pour down for weeks. While a rainy season is part of a monsoon, a monsoon is more than just rain. In fact, monsoons can also cause dry weather. Monsoons are caused by a change in the direction of the wind that happens when the seasons change. In fact, even the word monsoon comes from the Arabic word Mausam, which means "season."

WHAT CAUSES A MONSOON?

A monsoon is caused by a seasonal shift in the winds. The winds shift because the temperature of the land and the temperature of the water are different as seasons change. For example, at the beginning of summer, the land warms up faster than bodies of water. Monsoon winds always blow from cold to warm. In the summer, warm air rising off the land creates conditions that reverse the direction of the wind.

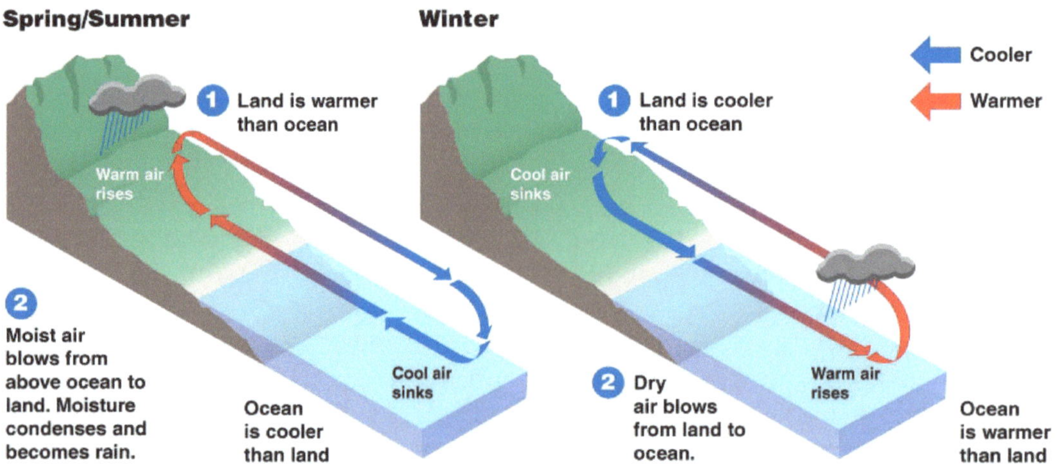

This diagram shows how seasonal temperature differences between the land and ocean can create the right conditions for a monsoon. Credit: NOAA/JPL

Source: https://scijinks.gov/what-is-a-monsoon/

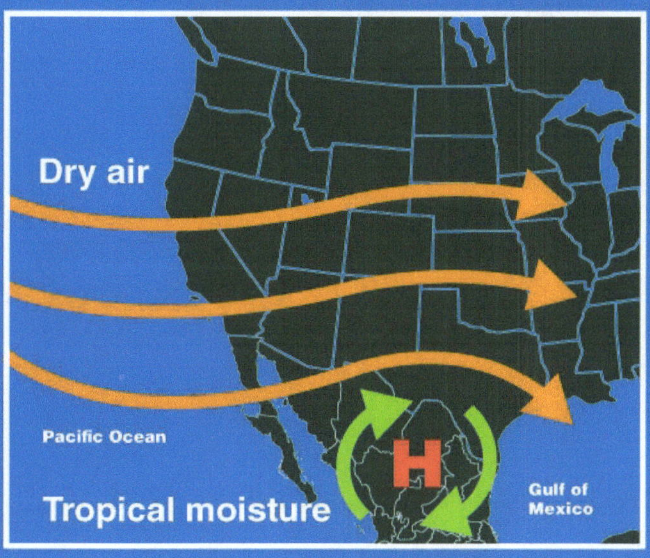

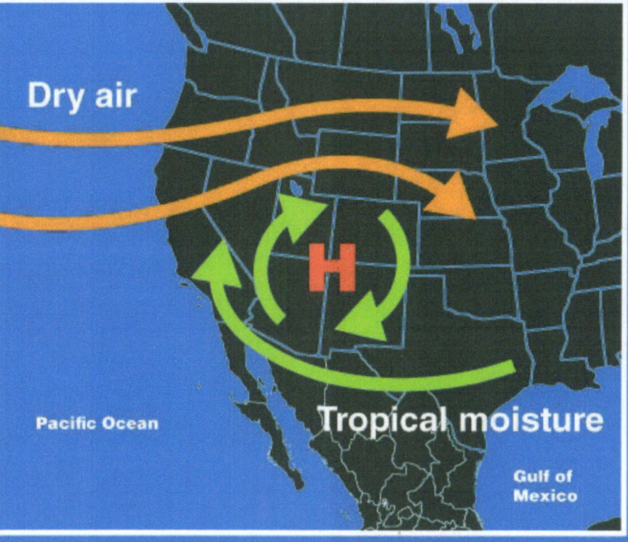

WHY DOES A MONSOON CAUSE RAIN?

The monsoons that cause heaviest rainfall are summer monsoons near the Indian Ocean. Warm water in the ocean evaporates, rising into the air. This causes the wind to change direction and moisture blows toward the land in countries such as India and Sri Lanka. The warm, moist air then condenses and becomes rain. The result is a period of humidity and heavy rainfall that can last for months.When the wind changes direction in the winter, it is called a winter monsoon. Winter monsoons in these regions near the Indian Ocean are usually dry.

WHERE DO MONSOONS HAPPEN?

While many of the most well-known monsoons are in Asia, monsoons can happen anywhere there is a seasonal difference in temperature between the land and water. This is usually in tropical and subtropical climates.The North American Monsoon is a seasonal change in wind that occurs as the summer sun heats the land of North America. During much of the year, the strongest winds over northwestern Mexico, Arizona and New Mexico are dry air blowing from the west.

As the land heats up in the summer, the wind begins to change direction and blows from the south. This new wind blows moist air from the Pacific Ocean and Gulf of California into the region, resulting in thunderstorms and rainfall.In general, the North American Monsoon begins each year in Mexico in June. The winds blowing from the south move the monsoon to the U.S. Southwest in July. By mid-September, winds are once again blowing from the west, marking the end of the monsoon.

During typical conditions in the spring, the U.S. Southwest experiences strong, dry winds blowing from the west. During a summer monsoon, the region experiences winds from the south, which carry moisture from the Pacific Ocean and the Gulf of California. This can cause heavy rainfall and thunderstorms ● Credit: NOAA/JPL

Source: https://scijinks.gov/what-is-a-monsoon/petrified-forest-national-park_monsoon.jpg
Source:https://scijinks.gov/what-is-a-monsoon/monsoon-conditions.jpg
Source: https://scijinks.gov/what-is-a-monsoon/typical-spring-pattern_and_summer-monsoon-pattern.jpg
Reference link: https://scijinks.gov/what-is-a-monsoon/

GROUNDWATER

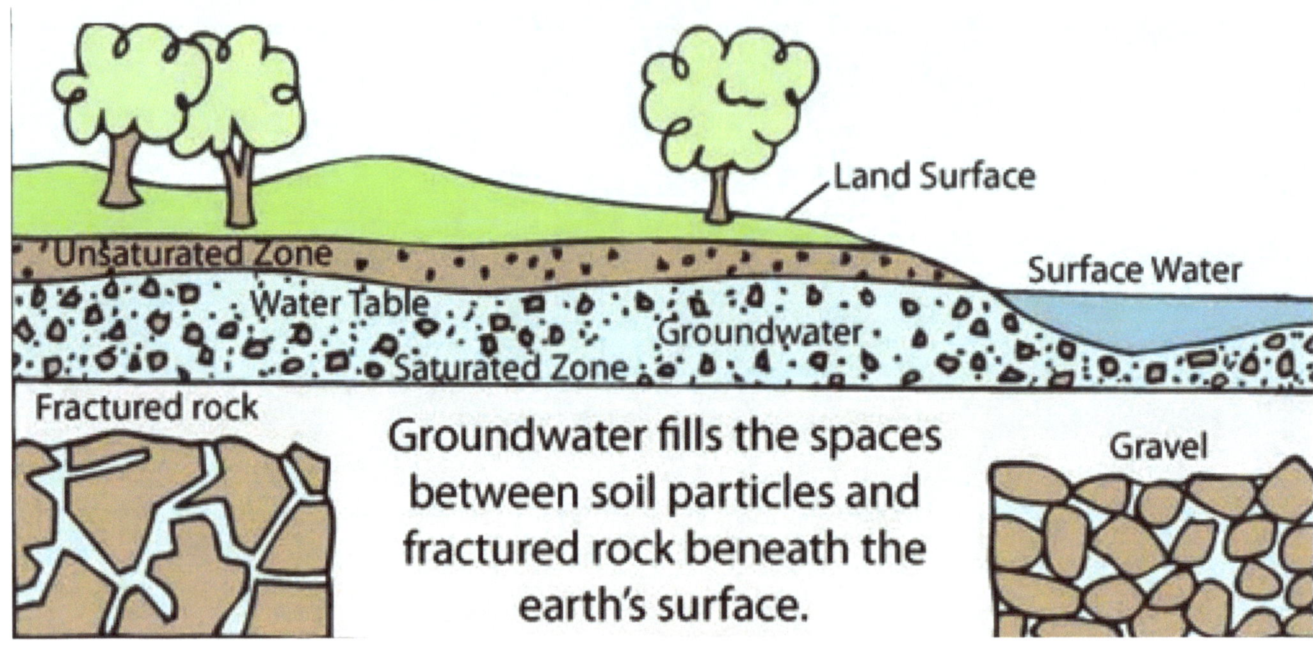

WHAT IS GROUNDWATER?

Groundwater is the water found underground in the cracks and spaces in soil, sand and rock. It is stored in and moves slowly through geologic formations of soil, sand and rocks called aquifers.Groundwater is used for drinking water by more than 50 percent of the people in the United States, including almost everyone who lives in rural areas. The largest use for groundwater is to irrigate crops. The area where water fills the aquifer is called the saturated zone (or saturation zone). The top of this zone is called the water table. The water table may be located only a foot below the ground's surface or it can sit hundreds of feet down. Aquifers are typically made up of gravel, sand, sandstone, or fractured rock, like limestone. Water can move through these materials because they have large connected spaces that make them permeable. The speed at which groundwater flows depends on the size of the spaces in the soil or rock and how well the spaces are connected. Groundwater can be found almost everywhere. The water table may be deep or shallow; and may rise or fall depending on many factors. Heavy rains or melting snow may cause the water table to rise, or heavy pumping of groundwater supplies may cause the water table to fall. Groundwater supplies are replenished, or recharged, by rain and snow melt that seep down into the cracks and crevices beneath the land's surface. In some areas of the world, people face serious water shortages because groundwater is used faster than it is naturally replenished. In other areas groundwater is polluted by human activities. Water in aquifers is brought to the surface naturally through a spring or can be discharged into lakes and streams. Groundwater can also be extracted through a well drilled into the aquifer. A well is a pipe in the ground that fills with groundwater. This water can be brought to the surface by a pump. Shallow wells may go dry if the water table falls below the bottom of the well. Some wells, called artesian wells, do not need a pump because of natural pressures that force the water up and out of the well. In areas where material above the aquifer is permeable, pollutants can readily sink into groundwater supplies. Groundwater can be polluted by landfills, septic tanks, leaky underground gas tanks, and from overuse of fertilizers and pesticides. If groundwater becomes polluted, it will no longer be safe to drink ●

HOW MUCH DO WE DEPEND ON GROUNDWATER?

Groundwater supplies drinking water for 51% of the total U.S. population and 99% of the rural population. Groundwater helps grow our food. 64% of groundwater is used for irrigation to grow crops. Groundwater is an important component in many industrial processes. Groundwater is a source of recharge for lakes, rivers, and wetlands.

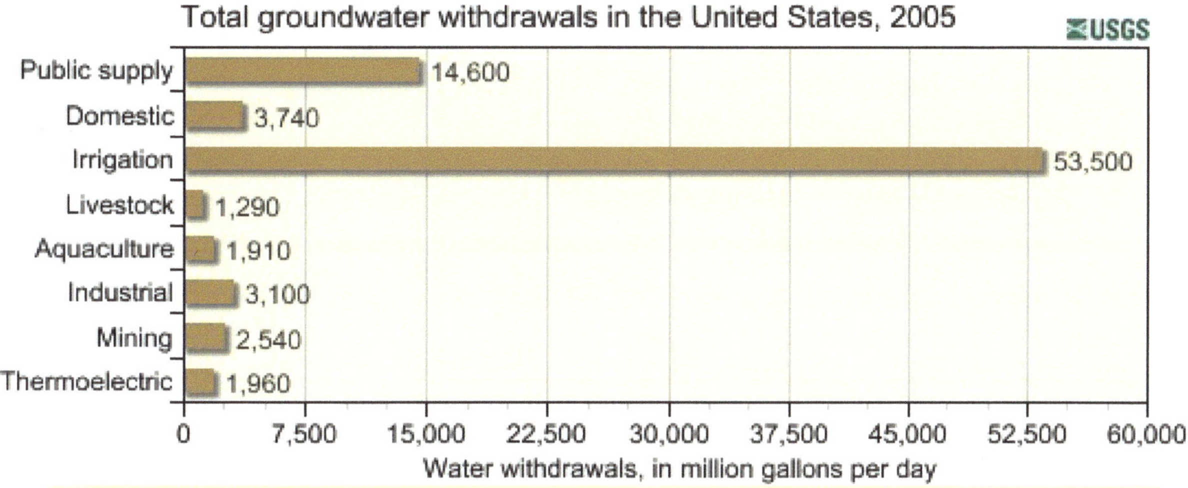

Image and figures courtesy of the U.S. Geological Survey

GROUNDWATER OVERUSE AND DEPLETION

Groundwater is the largest source of usable, fresh water in the world. In many parts of the world, especially where surface water supplies are not available, domestic, agricultural, and industrial water needs can only be met by using the water beneath the ground.

The U.S. Geological Survey compares the water stored in the ground to money kept in a bank account. If the money is withdrawn at a faster rate than new money is deposited, there will eventually be account-supply problems. Pumping water out of the ground at a faster rate than it is replenished over the long-term causes similar problems. Groundwater depletion is primarily caused by sustained groundwater pumping. Some of the negative effects of groundwater depletion:

Lowering of the Water Table. Excessive pumping can lower the groundwater table, and cause wells to no longer be able to reach groundwater.

Increased Costs. As the water table lowers, the water must be pumped farther to reach the surface, using more energy. In extreme cases, using such a well can be cost prohibitive.

Reduced Surface Water Supplies. Groundwater and surface water are connected. When groundwater is overused, the lakes, streams, and rivers connected to groundwater can also have their supply diminished.

Land Subsidence. Land subsidence occurs when there is a loss of support below ground. This is most often caused by human activities, mainly from the overuse of groundwater, when the soil collapses, compacts, and drops.

Water Quality Concerns. Excessive pumping in coastal areas can cause saltwater to move inland and upward, resulting in saltwater contamination of the water supply ●

SEA LEVEL RISE

WHAT IS SEA LEVEL RISE?

Sea level rise, as the name implies, is an increase in the total volume of ocean water. It results from the addition of melting glaciers and polar ice sheets, as well as the natural expansion of water as it warms—both consequences of climate change, which is driven by the burning of fossil fuels.

TYPES OF SEA LEVEL RISE

The sea isn't actually level. Because the earth itself is lumpy, the effect of gravity is uneven around the globe, subtly pulling more water into certain places than others. At the same time, the tides, ocean currents, and storms also cause ocean water to "pile up" in certain areas, leading to local variations in sea levels. Scientists who study sea level rise—using tide stations and satellite laser altimeters—try to account for all these complexities to hone in on long-term changes. There are two types of sea level rise:

1. GLOBAL MEAN SEA LEVEL RISE

Global mean sea level rise (also known as absolute sea level change) is the global average sea level compared to a fixed point, such as the center of the earth. This is the kind of sea level rise most related to climate change. Think of it as an increase in the total amount of space that the ocean takes up. (It may also be referred to as eustatic sea level rise.)

Global sea level rise has two primary causes: melting ice and the expansion of seawater as it warms. Climate change is causing both.

Melting ice : As climate change leads to an increase in the average temperatures at the earth's polar regions, the ice sheets are melting into the ocean and causing sea levels to rise. Meltwater from ice sheets in Greenland and Antarctica is a major driver of the global average rise in sea levels since at least 1993, with satellite data suggesting that the Greenland ice sheet alone sheds about 270 billions tons of ice mass every year.

Thermal expansion: Physics 101 says that liquids expand when heated. You can see thermal expansion at work in a mercury thermometer: As the temperature rises, the mercury expands, causing it to fill more of the glass tube. The oceans absorb 90 percent of the excess heat trapped by greenhouse gases. As climate change heats the oceans, the water gets bigger, even if the total amount of water were to remain constant. Since the ocean basin isn't getting bigger, the water level is pushed upward, just like the mercury in the thermometer. NOAA scientists have estimated that one-third of global sea level rise since 2004 is due to warming of the water.

2. RELATIVE SEA LEVEL CHANGE

Relative sea level change (also known as local or isostatic) describes the height of the ocean's surface compared to a specific piece of land. Increases may be caused by the water rising due to ocean dynamics, but it also may be that the land is sinking.

OCEAN WARMING

The ocean absorbs most of the excess heat from greenhouse gas emissions, leading to rising ocean temperatures.Increasing ocean temperatures affect marine species and ecosystems, causing coral bleaching and the loss of breeding grounds for marine fishes and mammals. Rising ocean temperatures also affect the benefits humans derive from the ocean; threatening food security, increasing the prevalence of diseases, causing more extreme weather events and the loss of coastal protection.

Achieving the mitigation targets set by the Paris Agreement on climate change and limiting the global average temperature increase to well below 2°C above pre-industrial levels is crucial to prevent the massive, irreversible impacts of ocean warming.Establishing marine protected areas and putting in place adaptive measures, such as precautionary catch limits to prevent over-fishing, can protect ocean ecosystems and shield humans from the effects of ocean warming ●

Source: https://i0.wp.com/apeejay.news/wp-content/up-loads/2024/03/210324-Rising-Sea-levels.jpg?resize=768%2C511&ssl=1 Reference link: https://www.nrdc.org/stories/sea-level-rise-101#what-is

MELTING GLACIERS

WHAT IS A GLACIER AND HOW DOES IT FORM?

These massive blocks of moving ice arise as snow accumulated in cold places compacts and recrystallizes, as is the case, for example, in mountain and polar glaciers, which should not be confused with the gigantic Arctic plates. Glaciers are classified according to their morphology — ice fields, cirque glaciers, valley glaciers, etc. — the climate — polar, tropical or temperate — and their thermal conditions — cold, hot or polythermal base —.

The formation of a glacier takes millennia, and its size varies depending on the amount of ice it retains throughout its lifespan. The behaviour of these masses is reminiscent of that of the rivers they feed during thaws, and their speed depends on friction and the slope of the terrain over which they move. In total, glaciers cover 10% of the Earth's surface and, along with the ice caps, account for nearly 70% of the world's fresh water.

WHY DO GLACIERS MELT? CAUSES

The rising temperature of the Earth has, without doubt, been responsible for melting glaciers throughout history. Today, the speed with which climate change is progressing might render them extinct in record time. Let us take a detailed look at the causes behind glacial melting:

CO_2 emissions: the atmospheric concentration of carbon dioxide and other greenhouse gases (GHGs) produced by industry, transport, deforestation and burning fossil fuels, amongst other human activities, warm the planet and cause glaciers to melt.

Ocean warming: oceans absorb 90% of the Earth's warmth, and this fact affects the melting of marine glaciers, which are mostly located near the poles and on the coasts of Alaska (United States).

EFFECTS OF MELTING GLACIERS

In the aforementioned study, the University of Zurich revealed that glacial melting has accelerated over the last three decades. This loss of ice has already reached 335 billion tonnes per year, which is 30% of the current rate of ocean growth. The main consequences of deglaciation are:

Sea level rise: Glacial melting has contributed to raising sea levels by 2.7 centimeters since 1961. Furthermore, the world's glaciers contain enough ice — about 170,000 cubic kilometres — to raise sea levels by nearly half a metre.

Impact on the climate: Glacial thawing at the poles is slowing the oceanic currents, a phenomenon related to altering the global climate and a succession of increasingly extreme weather events throughout the globe.

Disappearance of species: Glacial melting will also cause the extinction of numerous species, as glaciers are the natural habitat of a number of animals, both terrestrial and aquatic.

Less fresh water: The disappearance of glaciers also means less water for consumption by the population, a lower hydroelectric energy generation capacity, and less water available for irrigation.

SOLUTIONS TO AVOID MELTING GLACIERS

Glaciologists believe that, despite the massive ice loss, we do still have time to save the glaciers from their predicted disappearance. Here are some ideas and proposals for how we can help achieve this goal:

Stop climate change: In order to curtail climate change and save the glaciers, it is indispensable that global CO_2 emissions be reduced by 45 % over the next decade, and that they fall to zero after 2050. *Slow down their erosion:* The scientific journal Nature suggested building a 100-metre-long dam in front of the Jakobshavn glacier (Greenland), the worst affected by Arctic melting, to contain its erosion.

Combine artificial icebergs: Indonesian architect Faris Rajak Kotahatuhaha won an award for his project Refreeze the Arctic, which consists of collecting water from melted glaciers, desalinating it and refreezing it to create large hexagonal ice blocks. Thanks to their shape, these icebergs could then be combined to create frozen masses.

Increase their thickness: The University of Arizona proposed a seemingly simple solution: manufacture more ice. Their proposal consists of collecting ice from below the glacier through pumps driven by wind power to spread it over the upper ice caps, so that it will freeze, thus strengthening the consistency ●

Reference link: https://www.iberdrola.com/sustainability/melting-glaciers-causes-effects-solutions
Source: https://www.iberdrola.com/documents/20125/40372/glaciares_746x419.jpg/e509ee60-6ada-bbb6-3c2c-f489848781ce?t=1627294564901

LANDSLIDE

WHAT IS A LANDSLIDE?

A landslide is a mass movement of material, such as rock, earth or debris, down a slope. They can happen suddenly or more slowly over long periods of time. When the force of gravity acting on a slope exceeds the resisting forces of a slope, the slope will fail and a landslide occurs. External factors can lead to landslides happening, including:

- Heavy rainfall leading to saturation of the ground
- Erosion of the base of a slope
- Changes to the material's strength through weathering

Landslides are classified by their type of movement. The four main types of movement are:

- Falls
- Topples
- Slides (rotational and translational)
- Flows

Landslides can be classified as just one of these movements or, more commonly, can be a mixture of several.At BGS we usually study the landslides that happen in the UK, but here we also include some examples of large landslides that have occurred elsewhere around the world and also in the sea.

WHY DO LANDSLIDES HAPPEN?

A landslide may occur because the strength of the material is weakened. This reduces the power of the 'glue' that cements the rock or soil grains together. Located on a slope, the rock is then no longer strong enough to resist the forces of gravity acting upon it.

WHAT CAN INCREASE THE CHANCE OF A LANDSLIDE?

Several factors can increase a slope's susceptibility to a landslide event.

- *Water:* Adding water to the material on a slope makes a landslide more likely to happen. This is because water adds weight, lowers the strength of the material and reduces friction, making it easier for material to move downslope.

- *Erosion processes:* If the bottom of a slope is continually eroded, for example by the sea or a river, the slope will eventually become too steep to hold itself up.

- *Slope angle (steepness of slope):* The slope angle is a key factor as far as landslides are concerned. Any change to this that makes it steeper (such as coastal erosion) increases the likelihood of a landslide.

- *Rock type:* The type of rocks in the slope, and their combination, can increase the chance of a landslide.

- *Grain shape:* The shape of the grains that make up a rock can affect the risk of a landslide.

- Jointing and orientation of bedding planes.

- Arrangement of the rock layers.

- *Weathering processes:* For example, freeze-thaw reduces the cohesion ('stickiness') between the rock grains.

- *Vegetation:* Vegetation helps bind material together; removing vegetation increases the chance of a landslide.

- Flooding.

- Volcanoes and earthquake activity nearby.

- *Human activity:* mining, traffic vibrations or urbanisation change surface water drainage patterns.

HERE IS AN EXAMPLE OF LANDSLIDES OCCURRED IN KERALA

The Wayanad landslides in Kerala, India, were a tragic event that occurred on July 30, 2024. Heavy rainfall triggered landslides in several villages, resulting in significant loss of life and property. The affected areas, including Punjirimattom, Mundakkai, Chooralmala, and Vellarimala, suffered widespread destruction as entire villages were buried under debris. As of August 21, 2024, 231 bodies have been recovered, and 119 people remain missing. The landslides highlight the serious danger posed by landslides, especially in regions prone to heavy rainfall and geological instability. Such events can have devastating consequences for communities, leading to loss of lives, displacement, and economic hardship.

DESERTIFICATION

DEFINITION OF DESERTIFICATION

Desertification is the process by which vegetation in drylands i.e. arid and semi-arid lands, such as grasslands or shrublands, decreases and eventually disappears. The concept does not refer to the physical expansion of existing deserts, but to the various processes that threaten to turn currently non-desert ecosystems into deserts.

According to the UN, more than 24 billion tonnes of fertile soil disappear every year. In fact, today two-thirds of the Earth is undergoing a process of desertification and, if no action is taken, 1.5 million km2 of agricultural land, an area equivalent to the entire arable land of India, which is essential for maintaining biodiversity and feeding the population, will be lost by 2050.

DESERTIFICATION AND DECORTICATION: DIFFERENCES

Although they are often used interchangeably, the difference lies in the human influence on the process. In decortication, the causes of deterioration are strictly natural, as in the case of the Sahara mentioned above, but in desertification, although natural causes also play a role, human activities are a determining factor.

HOW TO AVOID DESERTIFICATION

Among the Sustainable Development Goals (SDGs) adopted by the UN is SDG 15 (Life of terrestrial ecosystems), which aims to protect, restore and promote the sustainable use of terrestrial ecosystems, sustainably manage forests, stop and reverse land degradation, combat desertification and stop biodiversity loss.

The solution at the local level to curb desertification is sustainable management of natural resources, especially the conservation of fertile soils and water resources. In this sense, some of the keys that can help to avoid desertification are:

● Promote coordinated land-use planning, including the management of water resources, livestock and agricultural activities.

● Preserve vegetation cover, which plays a key role in protecting the soil from wind and water erosion, by building barriers and stabilising dunes.

● Promote climate change education to raise awareness, in particular by showing the consequences of desertification and ways to prevent it.

● Focus on organic farming and sustainable practices, such as cover crops or rotational crops, which prevent soil erosion and drought.

● Commit to reforestation to regenerate vegetation cover, reactivate moisture circulation and generate biodiversity.

● Encourage rotational grazing, which limits pressure to a particular area while others regenerate, through co-existence with crops that allow more efficient nutrient cycling

CAUSES OF DESERTIFICATION

The main human activities driving desertification are:

● Deforestation, causes of which go beyond tree felling, which increases the risk of fires, among others. Poor agricultural practices, from not rotating crops to unprotected soils or chemical fertiliser and pesticide use, etc.
● Overexploitation of natural resources as a consequence, for example, of irresponsible management of vegetation or water. Bad livestock practices, such as overgrazing, which severely erode the land and prevent the regeneration of vegetation

CONSEQUENCES OF DESERTIFICATION

Drylands cover about half of the earth's ice-free land surface and many of them belong to the world's poorest countries, which exacerbates the consequences:
● Loss of biodiversity by worsening the living conditions of many species.
● Food insecurity due to crop failure or reduced yields.
● The loss of vegetation cover and therefore of food for livestock and humans.
● Increased risk of zoonotic diseases, such as COVID-19.
● Loss of forest cover, with a corresponding shortage of wood resources.

Reference link: https://www.iberdrola.com/sustainability/desertification#:~:text=Desertification%20is%20the%20process%20by,shrublands%2C%20decreases%20and%20eventually%20disappears
Source: https://media.tehrantimes.com/d/t/2022/04/29/4/4135403.jpg

BIODIVERSITY LOSS

WHAT IS BIODIVERSITY LOSS?

The definition of biodiversity loss is described as the loss of life on Earth at various levels, ranging from reductions in the genetic diversity to the collapse of entire ecosystems. In addition to its intrinsic value, biodiversity underpins ecosystem services, providing the backbone of the global economy.

CAUSES OF BIODIVERSITY LOSS

While biodiversity loss can occur naturally from more permanent ecological changes in ecosystems, landscapes, and the global biosphere, the current rapid rate of loss is a direct result of rampant human activity since the Industrial Revolution.

1. HABITAT LOSS

The single biggest contributor of global biodiversity loss is undoubtedly land and forest clearing to make way for urban and agricultural development. The latter of which has caused the loss of millions of hectares of trees to support the growing cattle and livestock industries, as well as mining activities. These ever-expanding industries are driven by global meat consumption, demand for commodities like paper and wood, and resources such as gold and other valuable minerals.

Decades of persistent land clearing means that we have lost significant amounts of natural habitats. Since forests especially are home to more than 80% of all terrestrial species of animals, plants and insects on the planet, millions of species have lost critical habitats to find shelter from prey and for reproduction, as well as increased food competition, causing population decline for many animals (and plants).

2. WILDLIFE TRADING

Animal poaching, wildlife and exotic pet trading have cost the lives of millions of animals from thousands of species across the world, causing nearly 30,000 species to become extinct every single year. Rare and vulnerable animal species are frequently targeted, caught and killed for food, as trophies, status symbols – for instance, elephant ivories and rhino horns, tourist ornaments, as well as allegedly medicinal purposes – many bears and tigers are killed for parts believed to be medicinal cures and even aphrodisiacs.

3. OVERFISHING

Aside from habitat loss from deforestation, another contributing factor in biodiversity loss is overfishing prompted by the commercial fishing industry. Today, we fish at a much higher and faster rate than fish stocks are able to replenish, pushing many fish species to the brink of extinction. While there are a number of regulations and fishing quotas in place to reduce the risk of overfishing – a few commercially-fished tuna species have recently been reported to show signs of population recovery – many other marine species including sharks and manta rays are in decline. This partly can be attributed to bycatching, where unwanted sea animals are captured during commercial fishing. About 38.5 million tonnes of bycatch result from unsustainable fishing practises every year.

4. CLIMATE CHANGE

Our dependence on fossil fuels and the resulting greenhouse gas emissions from it have created the phenomenon that is climate change. But today's climate is changing faste than species can move or adapt, and rising global temperatures are driving many animals to habitats they are not suited for. According to a 2004 study, scientists estimated that millions of species worldwide could face extinction as a result of climate changes predicted to occur in the next 50 years

EXAMPLE OF BIO-DIVERSITY LOSS

The Amazon Rainforest, often referred to as the "lungs of the Earth," is one of the most biodiverse regions in the world. It houses an estimated 10% of the known species on the planet, including countless plants, animals, and insects. However, large-scale deforestation, driven by activities such as logging, agriculture, and mining, is causing significant biodiversity loss.For instance, consider the Golden Lion Tamarin, a small, brightly coloured monkey native to the Atlantic Forest region of Brazil, which is part of the broader Amazon ecosystem. As deforestation progresses, its habitat is being fragmented and reduced. This loss of habitat directly threatens the species' survival, as the tamarins rely on specific forest conditions for food, shelter, and breeding.Additionally, deforestation disrupts the delicate ecological balance of the rainforest. Many species are interdependent, and the removal of trees affects not only the animals that live in them but also those that rely on the plants for food and shelter. The loss of large tracts of forest can lead to the decline of numerous species, some of which may become endangered or extinct if their habitat continues to disappear ●

SALINIZATION

WHAT IS SALINIZATION?

Salinization is the increase of salt concentration in soil and is, in most cases, caused by dissolved salts in the water supply. This supply of water can be caused by flooding of the land by seawater, seepage of seawater or brackish groundwater through the soil from below. Due to climate change, sea levels are rising, which further accelerates the process of salinization.

WHY IS SALINIZATION A PROBLEM?

Salinization of farmland is a fast-growing problem worldwide. Due to rising sea levels, vast areas of formerly arable land become increasingly saline.Since most farmers around the globe are unfamiliar with saline agriculture, their common belief is that saline soil is not suitable for growing crops. As a result, farmers feel forced to migrate and/or live with famine and poverty. Once arable land becomes a wasteland.

SALINIZATION A THREAT TO GLOBAL FOOD SECURITY

In today's world, climate change is causing issues such as drought and seawater levels to rise. As a result, farmlands are becoming more saline, making it difficult for smallholder farmers across the globe to produce food for their family, their cattle and for sale at markets. By using traditional farming practices for irrigation and fertilization, salinity levels have even been seen to increase. In many places, salt-affected lands become non-arable. In addition to the struggle farmers face, salinization is a serious threat to global food security. According to (FAO,2009), the world population is expected to grow to 9.1 billion people by 2050.

A FARMER'S PROBLEM TODAY, EVERYONE'S PROBLEM TOMORROW

Daily, farmers around the world suffer from salinization and are forced to live in poverty. And eventually, this will affect us all: as the world population grows, so does the demand for food. But the reality is that the amount of land suitable for agriculture continues to decrease. Due to salinization, vast areas of farmland are lost every day; land that is desperately needed to feed the world's growing population.

WHERE DOES SALINIZATION OCCUR?

Salinization is a global problem. It tends to be concentrated in the world's arid and semi-arid regions. The extent of problems related to salinity is predicted to increase, not only because of changes in the climate but also as a result of poor irrigation practices.

Effects of salinity on crops

Salt affects plants in several ways: there are an osmotic component and anionic component to salinity stress. Because of a large amount of dissolved salts in a salinized soil, it becomes difficult for a plant to take up sufficient water because the osmotic potential of the soil becomes very negative. Water will go from a place with less dissolved particles to a place with more dissolved particles (i.e. it will flow to the place with the lowest osmotic potential). This aspect of salinity problems in plants lead to 'physiological drought'. In this respect, salt stress is similar to drought stress.

The other component of salinity stress is related to the large amount of salts, usually NaCl (table salt) in the water which ends up in the plant. Na+ is toxic to plants. Besides this problem, the large amount of (positively charged) Na+ in the root zone makes it harder for a plant to absorb other positively charged ions such as potassium (K+) and calcium (Ca++) that the plant needs ●

Reference link: https://www.salineagricultureworldwide.com/salinization Source: https://eos.com/wp-content/uploads/2021/02/saline-soil-in-rice-field.jpg

CLOUD BURSTING

DEFINITION

A cloudburst is a sudden aggressive downpour within the radius of a couple of kilometres. Though, cloudbursts usually do not last for more than few minutes, they are capable of flooding the entire area. Rainfall from a cloudburst is usually equal to or greater than 100 mm per hour. Cloudbursts are generally associated with thunderstorms.However, the above definition has been given by a particular school of thought. In reality, there is no specific amount of rain associated with a cloudburst, either in time or duration.

HOW DOES A CLOUDBURST HAPPEN?

Whenever vertically formed clouds fully develop, moving very slow, most of the water content available in the clouds come down as a downpour at a particular area. Cloudbursts generally descend from very high clouds, sometimes with tops above 15 kilometers. The windward side of mountains are generally conducive for generating thunderclouds with huge updrafts.

EFFECT OF CLOUDBURSTS ON HILLS AND PLAINS

The catastrophic nature of cloudbursts differ on the virtue of terrain. In the hills, large volume of water keeps getting momentum as it flows in gushes. On its way, it demolishes everything and gravity of the situation increases due to landslides, mudslides, etc.On the other hand, cloudbursts in the plains only leads to waterlogging and inundation ●

Source: https://vajiramandravi.s3.us-east-1.amazonaws.com/media/2022/8/22/11/10/25/cloud.jpg
Reference link: https://www.skymetweather.com/content/weather-faqs/what-is-a-cloudburst/

WILD FIRE

A wildfire is an uncontrolled fire that burns in the wildland vegetation, often in rural areas. Wildfires can burn in forests, grasslands, savannas, and other ecosystems, and have been doing so for hundreds of millions of years. They are not limited to a particular continent or environment. Wildfires can burn in vegetation located both in and above the soil. Ground fires typically ignite in soil thick with organic matter that can feed the flames, like plant roots. Ground fires can smoulder for a long time—even an entire season—until conditions are right for them to grow to a surface or crown fire. Surface fires, on the other hand, burn in dead or dry vegetation that is lying or growing just above the ground. Parched grass or fallen leaves often fuel surface fires. Crown fires burn in the leaves and canopies of trees and shrubs.Some regions, like the mixed conifer forests of California's Sierra Nevada Mountain range, can be affected by different types of wildfires. Sierra Nevada forest fires often include both crown and surface spots.Wildfires can start with a natural occurrence—such as a lightning strike—or a human-made spark. However, it is often the weather conditions that determine how much a wildfire grows. Wind, high temperatures, and little rainfall can all leave trees, shrubs, fallen leaves, and limbs dried out and primed to fuel a fire. Topography plays a big part too: flames burn uphill faster than they burn downhill. Wildfires that burn near communities can become dangerous and even deadly if they grow out of control. For example, the 2018 Camp Fire in Butte County, California destroyed almost the entire town of Paradise; in total, 86 people died.Still, wildfires are essential to the continued survival of some plant species. For example, some tree cones need to be heated before they open and release their seeds; chaparral plants, which include manzanita, chamise (Adenostoma fasciculatum), and scrub oak (Quercus berberidifolia), require fire before seeds will germinate. The leaves of these plants include a flammable resin that feeds fire, helping the plants to propagate. Plants such as these depend on wildfires in order to pass through a regular life cycle. Some plants require fire every few years, while others require fire just a few times a century for the species to Continue.Wildfires also help keep ecosystems healthy. They can kill insects and diseases that harm trees. By clearing scrub and underbrush, fires can make way for new grasses, herbs, and shrubs that provide food and habitat for animals and birds. At a low intensity, flames can clean up debris and underbrush on the forest floor, add nutrients to the soil, and open up space to let sunlight through to the ground. That sunlight can nourish smaller plants and give larger trees room to grow and flourish.While many plants and animals need and benefit from wildfires, climate change has left some ecosystems more susceptible to flames, especially in the southwest United States. Warmer temperatures have intensified drought and dried out forests. The historic practice of putting out all fires also has caused an unnatural buildup of shrubs and debris, which can fuel larger and more intense blazes ●

AIR POLLUTION

WHAT IS AIR POLLUTION?

Air pollution refers to the release of pollutants into the air—pollutants that are detrimental to human health and the planet as a whole. According to the World Health Organization (WHO), each year, indoor and outdoor air pollution is responsible for nearly seven million deaths around the globe. Ninety-nine percent of human beings currently breathe air that exceeds the WHO's guideline limits for pollutants, with those living in low- and middle-income countries suffering the most. In the United States, the Clean Air Act, established in 1970, authorizes the U.S. Environmental Protection Agency (EPA) to safeguard public health by regulating the emissions of these harmful air pollutants.

WHAT CAUSES AIR POLLUTION?

"Most air pollution comes from energy use and production," says John Walke, director of the Clean Air team at NRDC. Driving a car on gasoline, heating a home with oil, running a power plant on fracked gas: In each case, a fossil fuel is burned and harmful chemicals and gases are released into the air. "We've made progress over the last 50 years in improving air quality in the United States, thanks to the Clean Air Act. But climate change will make it harder in the future to meet pollution standards, which are designed to protect health," says Walke.

EFFECTS OF AIR POLLUTION

The effects of air pollution on the human body vary, depending on the type of pollutant, the length and level of exposure, and other factors, including a person's individual health risks and the cumulative impacts of multiple pollutants or stressors.

Source: https:https://www.who.int/health-topics/air-pollution#:~:text=Air%20pollution%20is%20contamination%20of,common%20sources%20of%20air%20pollution. Reference link:https://www.nrdc.org/stories/air-pollution-everything-you-need-know#effects

SMOG AND SOOT

These are the two most prevalent types of air pollution. Smog (sometimes referred to as ground-level ozone) occurs when emissions from combusting fossil fuels react with sunlight. Soot—a type of particulate matter—is made up of tiny particles of chemicals, soil, smoke, dust, or allergens that are carried in the air. The sources of smog and soot are similar. "Both come from cars and trucks, factories, power plants, incinerators, engines, generally anything that combusts fossil fuels such as coal, gasoline, or natural gas," Walke says.Smog can irritate the eyes and throat and also damage the lungs, especially those of children, senior citizens, and people who work or exercise outdoors. It's even worse for people who have asthma or allegies; these extra pollutants can intensify their symptoms and trigger asthma attacks. The tiniest airborne particles in soot are especially dangerous because they can penetrate the lungs and bloodstream and worsen bronchitis, lead to heart attacks, and even hasten death. In 2020, a report from Harvard's T.H. Chan School of Public Health showed that COVID-19 mortality rates were higher in areas with more particulate matter pollution than in areas with even slightly less, showing a correlation between the virus's deadliness and long-term exposure to air pollution.

HAZARDOUS AIR POLLUTANTS

A number of air pollutants pose severe health risks and can sometimes be fatal, even in small amounts. Almost 200 of them are regulated by law; some of the most common are mercury, lead, dioxins, and benzene. "These are also most often emitted during gas or coal combustion, incineration, or—in the case of benzene—found in gasoline," Walke says. Benzene, classified as a carcinogen by the EPA, can cause eye, skin, and lung irritation in the short term and blood disorders in the long term. Dioxins, more typically found in food but also present in small amounts in the air, is another carcinogen that can affect the liver in the short term and harm the immune, nervous, and endocrine systems, as well as reproductive functions. Mercury attacks the central nervous system. In large amounts, lead can damage children's brains and kidneys, and even minimal exposure can affect children's IQ and ability to learn.

GREENHOUSE GASES

While these climate pollutants don't have the direct or immediate impacts on the human body associated with other air pollutants, like smog or hazardous chemicals, they are still harmful to our health. By trapping the earth's heat in the atmosphere, greenhouse gases lead to warmer temperatures, which in turn lead to the hallmarks of climate change: rising sea levels, more extreme weather, heat-related deaths, and the increased transmission of infectious diseases. In 2021, carbon dioxide accounted for roughly 79 percent of the country's total greenhouse gas emissions, and methane made up more than 11 percent. "Carbon dioxide comes from combusting fossil fuels, and methane comes from natural and industrial sources, including large amounts that are released during oil and gas drilling," Walke says. "We emit far larger amounts of carbon dioxide, but methane is significantly more potent, so it's also very destructive."

POLLEN AND MOLD

Mold and allergens from trees, weeds, and grass are also carried in the air, are exacerbated by climate change, and can be hazardous to health. Though they aren't regulated, they can be considered a form of air pollution. "When homes, schools, or businesses get water damage, mold can grow and produce allergenic airborne pollutants," says Kim Knowlton, professor of environmental health sciences at Columbia University and a former NRDC scientist. "Mold exposure can precipitate asthma attacks or an allergic response, and some molds can even produce toxins that would be dangerous for anyone to inhale."Pollen allergies are worsening because of climate change. "Lab and field studies are showing that pollen-producing plants—especially ragweed—grow larger and produce more pollen when you increase the amount of carbon dioxide that they grow in," Knowlton says. "Climate change also extends the pollen production season, and some studies are beginning to suggest that ragweed pollen itself might be becoming a more potent allergen."

FOOD INSECURITY

Food insecurity, the limited or uncertain access to nutritious food, which also includes limitations on the ability to obtain nutritious food in ways that are socially acceptable. Approximately 2.4 billion people worldwide (some 29.6 percent of the human population) experience moderate or severe food insecurity. Although food insecurity does occur in developed countries, the overwhelming majority of food-insecure people are concentrated in developing countries in Central and South America, Asia, and sub-Saharan Africa.

The Food and Agriculture Organization (FAO) of the United Nations (UN) divides food insecurity into two categories, moderate food insecurity (characterized by reduced food quality and quantity, the tendency to skip meals, and rising uncertainty about obtaining food) and severe food insecurity (characterized by running out of food and going without food for a day or more). Moderate food insecurity is a condition that affected some 1.5 billion people worldwide in 2022, and severe food insecurity impacted an additional 900 million people.

CAUSES OF FOOD INSECURITY

The causes of food insecurity are varied and complex and stem from a number of human-driven and natural factors. Almost three-quarters of people who do not have enough to eat live in politically unstable countries. War and other forms of conflict may affect people in the midst of the fighting, and it may also affect those outside the immediate region.

For example, the Russia-Ukraine War (2014–present) has reduced grain shipments from Ukraine to countries in sub-Saharan Africa that rely on them. Such supply-chain disruptions cause food shortages.

In addition, increases in fertilizer and fuel costs raise food prices, which reduces the amount of food poor individuals and countries can purchase. Climate change also contributes to food insecurity. Floods, droughts, and other changing weather conditions destroy crops and livestock while interfering with people's ability to work—further contributing to supply-chain disruptions.Food insecurity also occurs in wealthier countries.

For example, in early 2023 roughly one in ten U.S. adults were food insecure because of any of several factors, which include having low income or lack of income, rising costs associated with housing and medicine, the inability to access the medical system, systemic racism and discrimination, and the creation of food deserts.

The period of rising inflation in the U.S. that followed the COVID-19 pandemic combined with stagnant wages to result in lower discretionary income, which contributed to greater food insecurity. Food insecurity in wealthy countries can also be triggered by loss of employment, injury, unanticipated home repairs, and changes to government benefits that restrict and reduce the types of foods that can be purchased.

HEALTH EFFECTS

Food insecurity also contributes to poor health, since people may skip meals, eat less, or switch to lower-quality foods. Malnutrition can lead to weakness, pain, illness, and even death. Food insecurity also contributes to type 2 diabetes, obesity, and hypertension (high blood pressure) when less-expensive foods with lower nutritional value are substituted for healthy foods over the long term. Existing health issues can be worsened if people forgo necessary medication or medical treatment in order to buy food. Children who live in food-insecure homes may have difficulty learning and may not develop properly. Food insecurity is associated with poor mental health, low educational attainment, and poor job performance.

Source: https://www.voj.news/wp-content/uploads/2017/08/Food-Insecurity.jpg
Reference link: https://www.britannica.com/topic/food-insecurity

SOLUTIONS TO FOOD INSECURITY

Nearly one billion people experience hunger; however, there is enough food produced each year to feed the world's population. Food insecurity stems from challenges to food availability, food access, and food utilization. Food availability involves ensuring not only that enough food is available to feed a population but that food supplies are available consistently.

Food access, in contrast, involves having the resources necessary that give people the ability to obtain appropriate foods for a nutritious diet. Broadly speaking, food can be made more available by increasing a population's access to the resources that grow food (that is, fertilizer, water, and arable land) and to networks of food distribution.

In addition, governments, nongovernmental organizations (NGOs), and relief groups can improve the structures that provide people with jobs and other financial resources that help people purchase food, as well as helping them with food utilization—that is, the process of consuming a nutritious diet by having access to clean water, adequate sanitation, and the education necessary to both grow nutritious food (rather than low-nutrition food) and prepare it hygienically. To attain these goals worldwide, international organizations have partnered with local governments and other groups. In 2012 the UN launched the Zero Hunger Challenge with the goal of eradicating hunger by 2030, as part of its global sustainable development objectives. In 2021 the UN held a summit to address hunger and consider the progress made by the Zero Hunger Challenge effort.

While several UN organizations (including the World Hunger Programme and the FAO), lending organizations (such as the World Bank), and relief groups (such as the Peace Corps) continue to work with communities to prevent food insecurity, summit participants and organizers and a follow-on report ("The State of Food Security and Nutrition in the World," produced by the FAO) revealed that hunger had increased through the late 2010s and early 2020s because of the effects of climate change, conflict, and the distribution challenges brought on by the COVID-19 pandemic. They recommended that governments and NGOs work harder to improve sustainable food production and distribution systems through, for example, better communication between local and regional stakeholders, better management of water resources and fisheries, and accelerated financing to fund research, innovation, and the rollout of food-production projects.

WATER SCARCITY

Water scarcity is an increasing problem on every continent, with poorer communities most badly affected. To build resilience against climate change and to serve a growing population, an integrated and inclusive approach must be taken to managing this finite resource.

THE ISSUE EXPLAINED

Water scarcity is a relative concept. The amount of water that can be physically accessed varies as supply and demand changes. Water scarcity intensifies as demand increases and/or as water supply is affected by decreasing quantity or quality.

Water is a finite resource in growing demand. As the global population increases, and resource-intensive economic development continues, many countries' water resources and infrastructure are failing to meet accelerating demand.

Climate change is making water scarcity worse. The impacts of a changing climate are making water more unpredictable. Terrestrial water storage – the water held in soil, snow and ice – is diminishing. This results in increased water scarcity, which disrupts societal activity.

Women and girls are among the hardest hit. Poor and marginalized groups are on the frontline of any water scarcity crisis, impacting their ability to maintain good health, protect their families and earn a living. For many women and girls, water scarcity means more laborious, time-consuming water collection, putting them at increased risk of attack and often precluding them from education or work.

Lack of data means lack of integrated management. Many countries do not have well developed water monitoring systems, which prevents integrated water resource management that can balance the needs of communities and the wider economy, particularly in time of scarcity.

THE WAY FORWARD

Water has to be treated as a scarce resource. Integrated water resources management (IWRM) provides a broad framework for governments to align water use patterns with the needs and demands of different users, including the environment.

IWRM can control water stress. When a territory withdraws 25 per cent or more of its renewable freshwater resources it is said to be 'water-stressed'. IWRM can control water stress by measures such as reducing losses from water distribution systems, safe wastewater reuse, desalination and appropriate water allocation.

Data, technology and communications have a critical role. IWRM depends on: good quality data on water resources; water-saving, green and hybrid technologies, particularly in industry and agriculture; and awareness campaigns to reduce the use of water in households and encourage sustainable diets and consumption.

Groundwater is part of the solution. Exploring, protecting and sustainably using groundwater will be central to surviving and adapting to climate change and meeting the needs of a growing population.

Source: https://www.campdenfb.com/sites/campdendrupal.modezero.net/files/iStock-804500172%20Squeeze%20water%20Earth%20scarcity.jpg
Reference link: https://www.unwater.org/water-facts/water-scarcity

POPULATION DISPLACEMENT

The displacement of human populations refers to the relocation of large numbers of people from their homes. On this website, we focus on human displacement driven directly or indirectly by non-human or human-caused changes in the physical environment, which create shortages of essential resources. For example, climate change can reduce regional precipitation and convert farmland into desert. Unfortunately, the displacement of human populations is not only caused by environmental changes, but often occurs because of inequitable social and political systems that do not provide people with protection, food, clean water, and healthcare. Either sudden or long-term changes in Earth's environment can force or motivate people to leave their homes to preserve their health and well-being. These changes are caused by a variety of factors related to human use of Earth's resources, climate change, and non-human processes.

SOME OF THE REASONS FOR THE DISPLACEMENT OF HUMAN POPULATIONS INCLUDE:

- Decreases in the availability of freshwater for both drinking and agriculture.

- Decreases in quality, availability, and nutritional value of food.

- Extreme weather events, including droughts when precipitation decreases, hurricanes that bring high winds and flooding.

- Flooding due to sea level rise.

- Habitat loss from activities such as deforestation.

- Earthquakes, volcanic eruptions, or fires that alter landscapes.

CIRCULAR ECONOMY

WHAT IS CIRCULAR ECONOMY?

The concept of circular economy has gained traction in recent years, inspiring environmentalists, governments and businesses alike. Once a fringe topic, circularity is now acknowledged globally as the most promising solution to our planet's looming sustainability issues. However, many diverging definitions and understandings of circular economy exist.

A common misunderstanding minimizes the definition of circular economy to the familiar Reduce-Reuse-Recycle approach. But as Ellen MacArthur Foundation CEO Andrew Morlet explained during a Leading Disruption Panel in 2020: "Recycling alone will not save us." Circular economy is a "bigger idea" — a significant restructuring that forces us to rethink how we've done things since the rise of the first steam engine.

HOW DOES CIRCULAR ECONOMY WORK?

DESIGNS OUT WASTE AND POLLUTION

Circular economy designs out economic activities that negatively impact human health and natural systems. This includes the release of greenhouse gases, all types of pollution and traffic congestion.

KEEPS PRODUCTS AND MATERIALS IN USE

Circular economy favors designing products for durability, reuse, remanufacturing, and recycling to keep materials circulating for as long as possible. It's an economy that encourages many different uses for materials instead of just using them up.

REGENERATES LIVING SYSTEMS

Circular economy avoids the use of fossil fuels and non-renewable energy. By preserving and enhancing renewable resources, it returns valuable nutrients to the soil to support regeneration and actively improve the environment.

WHY IS CIRCULAR ECONOMY IMPORTANT?

Circular economy aims to throw away nothing, thereby reducing the need to use more commodities. It offers a stark alternative to our linear "take make-dispose" economy — an economy that runs on the assumption that there will always be virgin materials to turn into products, and always somewhere to put the waste.

As the world's population continues to grow, it's becoming increasingly clear the assumptions of linear economy aren't true or at the very least, sustainable. The model that has dominated manufacturing since the First Industrial Revolution has come under strain.

At that time less than 1 billion people inhabited the Earth. Today, the world's population is up to 8 billion — with a growing global middle class of consumers. Not only are we using the same resources, we're throwing them away at an alarming rate. According to a report from the United Nations, global resource extraction has more than tripled since 1970. Over 90% of raw materials are not reused.

Reference link: https://www.dotmagazine.online/issues/building-the-internet-of-tomorrow/future-of-payments-wallets/circular-economy-shaping-future

INTERGOVERNMENTAL PANEL ON CLIMATE CHANGE

The Intergovernmental Panel on Climate Change (IPCC) is the international body for assessing the science related to climate Change. The IPCC was set up in 1988 by the World Meteorological Organization (WMO) and United Nations Environment Programme (UNEP) to provide policymakers with regular assessments of the scientific basis of climate change, its impacts and future risks, and options for adaptation and mitigation.IPCC assessments provide a scientific basis for governments at all levels to develop climate related policies, and they underlie Negotiations at the UN Climate Conference – the United Nations Framework Convention on Climate Change (UNFCCC). The Assessments are policyrelevant but not policy prescriptive: they may present projections of future climate change based on Different scenarios and the risks that climate change poses and discuss the implications of response options, but they do not tell policymakers what actions to take. The IPCC is currently in its seventh assessment cycle which began in July 2023.The IPCC is a unique interface between science and policy. Because of its scientific and intergovernmental nature, its Assessments provide rigorous and balanced scientific information to decision-makers. Participation in the IPCC is open to all Member countries of the WMO and United Nations. It currently has 195 members. The Panel, made up of representatives of the member states, meets in Plenary Sessions to take major decisions. The IPCC Bureau, elected by member governments,provides guidance to the Panel on the scientific and technical aspects of the Panel's work and advises the Panel on related Management and strategic issues. IPCC assessments are written by hundreds of leading scientists who volunteer their time and expertise as Coordinating Lead Authors and Lead Authors of the reports. They enlist hundreds of other experts as Contributing Authors to provide Complementary expertise in specific areas. The authors may work with Chapter Scientists who cross-check between findings Presented in different parts of the report, carry out additional fact-checking, and work on reference management among other things. Chapter Scientists are usually early career scientists.How does the IPCC approve Reports?The IPCC reports go through multiple drafts and reviews before being finalized. Once completed, the reports are endorsed by IPCC member governments, ensuring that they follow the IPCC's procedures and hold authority. There are three levels of endorsement:

1.	Approval: Used for Summaries for Policymakers (SPMs), this involves detailed, line-by-line discussions between governments and scientists to ensure the SPM is clear and accurate.

2.	Adoption: Applied to Synthesis Reports, this involves section-by-section discussions to ensure the report effectively integrates material from other reports. The SPM of a Synthesis Report is also approved line by line.

3.	Acceptance: Used for the full underlying reports in Working Group or Special Reports. It signifies that the report is comprehensive and balanced but doesn't involve detailed line-by-line discussions. Changes after acceptance are limited to consistency adjustments with the SPM.
llenge.

Reference link: https://www.ipcc.ch/about/ https://www.ipcc.ch/

5R's (REFUSE, REDUCE, REUSE, REPURPOSE, RECYCLE)

According to the 5 R's, four actions should be taken, if possible, prior to 'recycling': refuse, reduce, reuse, repurpose, and then recycle. Incorporating this methodology into your business' waste reduction and recycling efforts will minimize landfill waste and help take your recycling program to the next level, helping you enter the world of sustainable waste management.

HOW TO APPLY THE 5 R'S

Applying the 5R's to your business' waste management and recycling strategies can positively impact the outcome of your program by significantly reducing the amount of waste your business generates. In the 5 R's hierarchy, remember to treat recycling as a last resort after

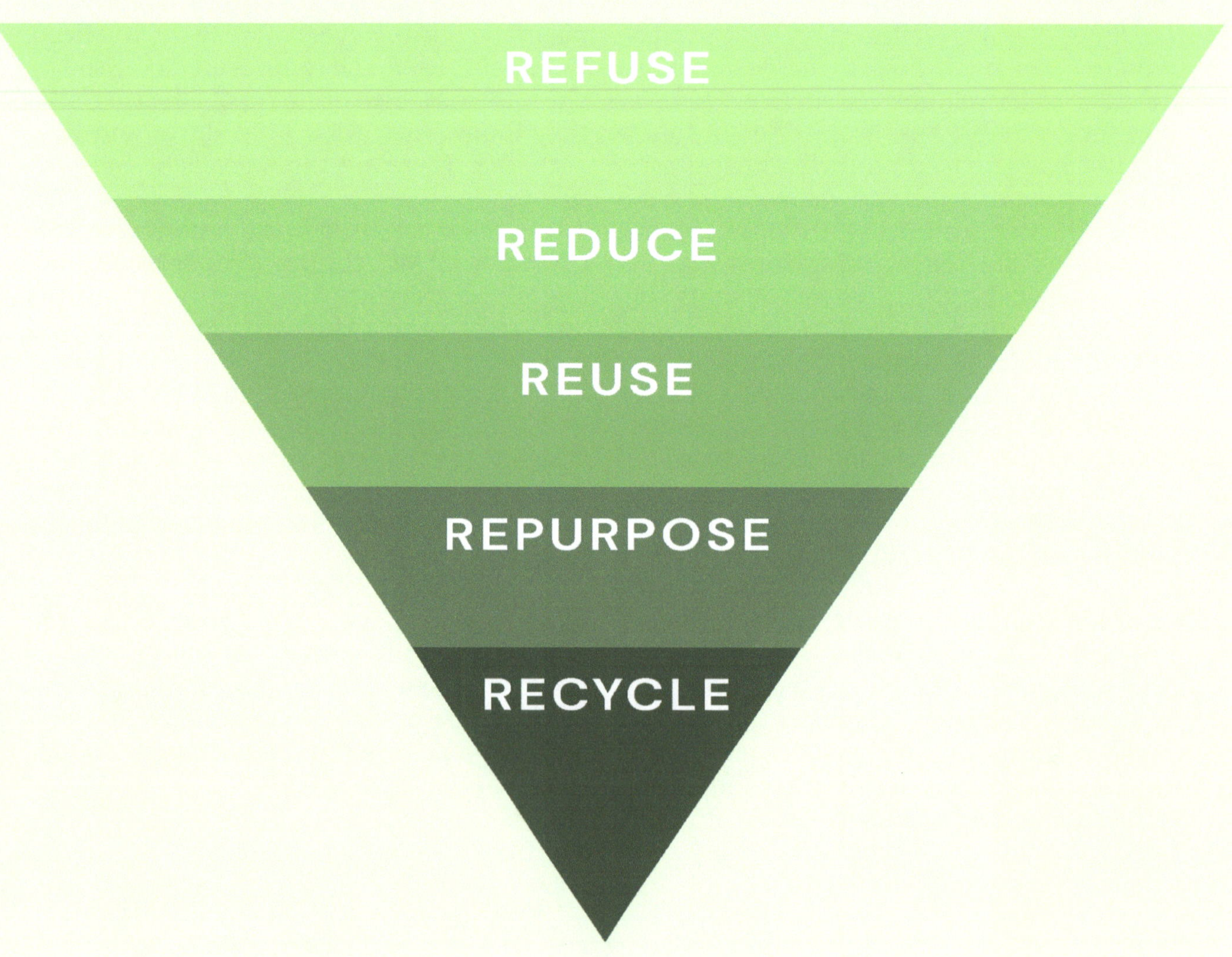

attempting to refuse, reduce, reuse or repurpose. Before disposing of your waste, walk through each of these steps in the following order:

STEP ONE: REFUSE

REFUSE: the first element of the 5 R's hierarchy. Learning to refuse waste can take some practice, but incorporating this step into your business' strategy is the most effective way to minimize waste. Talk to your procurement team about refusing to buy wasteful or non-recyclable products. When working with vendors, refuse unnecessary product packaging and request reusable or returnable containers.

STEP TWO: REDUCE

Reduce the use of harmful, wasteful, and non-recyclable products. Reducing dependency on these kinds of products results in less waste materials ending up in landfill and the associated negative environmental impacts. We recommend always using the minimum amount required to avoid unnecessary waste. For example, when printing a document, print double-sided to cut your waste output in half. Other commonly used items businesses can focus on reducing include single-use plastics, plastic packaging, organic waste, and Styrofoam cups.

STEP THREE: REUSE

Single-use plastics have created a "throw-away" culture by normalizing consumer behavior of using materials once and then throwing them away. The rate at which we consume plastics has become unimaginable, and the plastic crisis has become one of the world's greatest environmental challenges. In an effort to reduce waste, reuse items throughout the workplace instead of buying new ones. Begin by focusing on one area of your business at a time, like the break room. Replace all of the single-use eating utensils, Styrofoam cups, water bottles, and paper plates with compostable or reusable alternatives.

STEP FOUR: REPURPOSE

For every item that can't be refused, reduced, or reused, try repurposing it. Many people in the green community refer to this method as upcycling. You may be surprised to learn how many common office products serve more than one purpose. Sometimes it requires using some creativity, but the possibilities are endless.

STEP FIVE: RECYCLE

For every item that can't be refused, reduced, or reused, try repurposing it. Many people in the green community refer to this method as upcycling. You may be surprised to learn how many common office products serve more than one purpose. Sometimes it requires using some creativity, but the possibilities are endless.

WORLD TOP 20 CARBON EMITTING NATIONS
Here are the top 20 carbon-emitting nations:

1. CHINA:
12,667,428,430 tons of CO_2 emissions, accounting for 32.88% of the world's total

2. UNITED STATES:
4,853,780,240 tons, accounting for 12.60%

3. INDIA:
2,693,034,100 tons, accounting for 6.99%

4. RUSSIA:
1,909,039,310 tons, accounting for 4.96%

5. JAPAN:
1,082,645,430 tons, accounting for 2.81%

6. INDONESIA:
692,236,110 tons, accounting for 1.80%

7. IRAN:
686,415,730 tons, accounting for 1.78%

8. GERMANY:
673,595,260 tons, accounting for 1.75%

9. SOUTH KOREA:
635,502,970 tons, accounting for 1.65%

10. SAUDI ARABIA:
607,907,500 tons, accounting for 1.58%

11. CANADA:

582,072,950 tons, accounting for 1.51%

12. MEXICO:

487,774,010 TONS, ACCOUNTING FOR 1.27%

13. TURKEY:

481,247,520 TONS, ACCOUNTING FOR 1.25%

14. BRAZIL:

466,770,410 tons, accounting for 1.21%

15. SOUTH AFRICA:

404,974,510 tons, accounting for 1.05%

16. AUSTRALIA:

393,162,550 tons, accounting for 1.02%

17. UNITED KINGDOM:

340,610,260 tons, accounting for 0.88%

18. VIETNAM:

327,905,620 tons, accounting for 0.85%

19. ITALY:

322,948,740 tons, accounting for 0.84%

20. POLAND:

321,954,000 tons, accounting for 0.84%

13

Climate Action

SDG 13 CLIMATE ACTION

Sustainable Development Goals (SDG) 13 focuses on taking urgent action to combat climate change and its impacts. Climate records were shattered in 2023, with the world watching the climate crisis unfold in real time. Communities around the world are suffering the effects of extreme weather, which is destroying lives and livelihoods on a daily basis. The roadmap to limit the rise in global temperature to 1.5°C and avoid the worst of climate chaos cannot afford any delays, indecision or half measures by the global community. It demands immediate action for drastic reductions in global greenhouse gas emissions in this decade and the achievement of net zero by 2050.

Reference link: https://sdgs.un.org/goals/goal13#progress_and_info Source: https://ourworldindata.org/images/published/Thumbnail-SDG-13.png

Progress: The number of disaster-related deaths per 100,000 has nearly halved from 1.62 (2005-2014) to 0.82 (2013-2022), but the total annual deaths remain high at approximately 42,553.
Vulnerability: The number of people affected by disasters per 100,000 increased from 1,169 to 1,980 during the same periods, indicating rising vulnerability.

Record Temperatures: 2023 is the warmest year on record, with global temperatures reaching 1.45°C, nearing the 1.5°C limit set by the Paris Agreement.
Greenhouse Gas Levels: Despite some reductions in developed countries, greenhouse gas concentrations reached record highs in 2022 and continue to rise, with CO_2 levels 150% above pre-industrial levels.

Curriculum Gaps: A study found that 69% of grade 9 science and social science curricula lack references to climate change, and 66% do not mention sustainability.
Future Plans: Three-quarters of countries plan to revise their curricula in the next three years to better incorporate climate change and sustainability topics.

Funding Growth: Climate finance from developed countries to developing nations grew at a compound rate of 5% from 2015 to 2020, reaching $41 billion.
Funding Challenges: The goal of $100 billion annually was not met as of 2021, but mobilization efforts increased, reaching $89.6 billion in 2021.

GLASGOW FINANCIAL
ALLIANCE FOR NET ZERO (GFANZ)

The Glasgow Financial Alliance for Net Zero is the world's largest coalition of financial institutions committed to transitioning the global economy to net-zero greenhouse gas emissions.

HISTORY

GFANZ was launched in April 2021 by UN Special Envoy on Climate Action and Finance Mark Carney and the COP26 presidency, in partnership with the UNFCCC Race to Zero campaign, to coordinate efforts across all sectors of the financial system to accelerate the transition to a net-zero global economy. UN Special Envoy for Climate Ambition and Solutions Michael R. Bloomberg joined Mark Carney as Co-Chair at COP26 and brought in former SEC Chairman and Head of the Secretariat for the Task Force on Climate-related Financial Disclosures Mary Schapiro to serve as Vice Chair and Head of the GFANZ Secretariat.

WORK

GFANZ provides financial firms with a global forum for collaborating on substantive, cross-cutting issues that will accelerate the alignment of financing activities with net zero. Their work focuses on three core areas, each one critical to the net-zero transition:

Net-zero transition planning for financial institutions

Mobilizing capital for emerging markets and developed economies

Net-zero public policy

Reference link: https://www.gfanzero.com/about/
Source: https://imengine.public.prod.wbm.infomaker.io/?uuid=484dd721-9a59-5959-b3db-bc56d2f6bd9e&type=preview&source=false&q=75&width=1098

CARBON PRICING REVENUE

Carbon pricing revenue is the money collected by governments or organizations through various carbon pricing mechanisms. These mechanisms aim to reduce greenhouse gas emissions by placing a cost on them

COMMON CARBON PRICING MECHANISMS:

Carbon Taxes: A direct tax levied on the carbon content of fossil fuels or emissions.
Emissions Trading Systems (ETS): A market-based system where companies can buy and sell permits to emit a certain amount of greenhouse gases.

HOW CARBON PRICING REVENUE IS USED:

The revenue generated from carbon pricing can be used in several ways:

CLIMATE INVESTMENT	PUBLIC BENEFITS
• Funding research and development of clean technologies • Supporting renewable energy projects • Investing in energy efficiency programs	• Reducing taxes on other economic activities • Investing in public services like education and healthcare • Providing financial assistance to vulnerable populations affected by climate change.

Source: https://th.bing.com/th/id/OIP.XGHcS9-kgGWgZsz5xOOnaAAAAA?rs=1&pid=ImgDetMain   Reference link: https://gemini.google.com/

CARBON CREDIT
(QUALITY AND INTEGRITY)

Carbon credits are a market-based tool used to reduce greenhouse gas emissions. They represent a verified reduction or removal of one tonne of carbon dioxide equivalent (tCO_2e) from the atmosphere. The quality and integrity of these credits are crucial to ensure they effectively contribute to climate mitigation efforts.

KEY FACTORS AFFECTING CARBON CREDIT QUALITY:

● *Additionality:* The project must generate emissions reductions that would not have occurred without the carbon credit initiative.

● *Permanence:* The emissions reductions must be long-lasting and not likely to be reversed.

● *Measurability:* The quantity of emissions reductions must be accurately measured and verified.

● *Verifiability:* An independent third party should verify the project's claims of emissions reductions.

● *Leakage:* The project should not lead to increased emissions elsewhere (e.g., companies relocating to avoid regulations).

● *Double Counting:* The same emissions reductions should not be claimed by multiple parties.

ENSURING CARBON CREDIT INTEGRITY:

● *Standards and Certifications:* Adherence to recognized standards like Verra, Gold Standard, or the American Carbon Registry can help ensure quality.

Independent Verification: Third-party verifiers assess project activities and claims.

● *Transparency:* Clear reporting and disclosure of project information are essential.

● *Registry Systems:* Centralized registries track the issuance, transfer, and retirement of carbon credits.

● *Market Mechanisms:* Well-functioning carbon markets can incentivize the creation of high-quality credits.

CHALLENGES AND CONSIDERATIONS:

● *Complexity:* The carbon credit market can be complex, making it difficult for buyers to assess quality.

● *Risk of Reversal:* Some projects may face risks of emissions reversals, undermining the long-term impact of credits.

● *Social and Environmental Impacts:* Carbon credit projects should consider their social and environmental impacts, ensuring they contribute positively to sustainable development.

By prioritizing quality and integrity, carbon credits can play a valuable role in addressing climate change while promoting sustainable development and economic growth.

CARBON CREDIT

(INTERNATIONAL CREDITING MECHANISM/INDEPENDENT CREDITING MECHANISM/ GOVERNMENTAL CREDITING MECHANISM)

The international, independent, and governmental crediting mechanisms in carbon credit are different systems for verifying and certifying greenhouse gas emissions reductions. Each mechanism has its own specific rules, standards, and procedures, but they all ultimately serve the same purpose: to ensure that carbon credits represent real and measurable reductions in emissions.

INTERNATIONAL CREDITING MECHANISM (ICM):

- Established by the Kyoto Protocol.
- Primarily focuses on emissions reductions in developing countries.
- Utilizes Clean Development Mechanism (CDM) and Joint Implementation (JI) projects.
- CDM projects involve developed countries investing in emission reduction projects in developing countries.
- JI projects involve joint emissions reduction activities between developed countries.
- ICM credits are issued by the United Nations Framework Convention on Climate Change (UNFCCC).

INTERNATIONAL CREDITING MECHANISM (ICM):

- Independent of government or international organizations.
- Often established by non-governmental organizations (NGOs) or private companies.
- Sets its own standards and verification procedures.
- Issues credits based on its own criteria.
- Examples include Verified Carbon Standard (VCS) and Gold Standard.

GOVERNMENTAL CREDITING MECHANISM:

- Established and regulated by a government or its regulatory body.
- Often used for domestic emissions reduction programs.
- Sets its own standards and verification procedures.
- Issues credits based on its own criteria.
- Examples include the Australian Carbon Farming Initiative and the European Union Emissions Trading System (EU ETS).

Reference link: https://gemini.google.com/

CARBON CREDIT MARKET SEGMENTS

(DOMESTIC COMPLIANCE MARKETS/ INTERNATIONAL COMPLIANCE MARKETS/ RESULT-BASED FINANCE/ VOLUNTARY CARBON MARKETS)

The carbon credit market can be segmented based on various factors, including:

1. GEOGRAPHIC LOCATION

Domestic Markets: These are markets within a specific country or region, often regulated by local governments. For example, the Regional Greenhouse Gas Initiative (RGGI) in the northeastern United States.

International Markets: These markets operate across national borders, facilitated by international agreements like the Kyoto Protocol and the Paris Agreement. The Clean Development Mechanism (CDM) and Joint Implementation (JI) are examples of international mechanisms.

2. REGULATORY STATUS

Compliance Markets: These markets are established under mandatory regulatory frameworks, requiring participants to purchase or trade carbon credits to meet emissions reduction targets.

Voluntary Markets: These markets are not regulated by law and are driven by voluntary participation. Individuals, corporations, and organizations can purchase carbon credits to offset their emissions.

3. PROJECT TYPE

Renewable Energy: Credits generated from projects that produce renewable energy, such as solar, wind, and hydropower.

Forestry: Credits derived from reforestation, afforestation, and REDD+ (Reducing Emissions from Deforestation and Forest Degradation) projects.

Energy Efficiency: Credits generated from projects that improve energy efficiency, such as building retrofits or industrial process improvements.

Agriculture: Credits from projects that reduce agricultural emissions, such as methane emissions from livestock.

4. CREDIT STANDARD

Verified Carbon Standard (VCS): One of the most widely used standards for verifying carbon credits.

Gold Standard: A stringent standard that requires additional social and environmental safeguards.

American Carbon Registry (ACR): A U.S.-based standard.

5. CREDIT QUALITY

High-Quality Credits: Credits from projects that have rigorous verification processes and deliver additional benefits beyond emissions reduction, such as biodiversity conservation or community development.
Low-Quality Credits: Credits from projects with weaker verification or that may have questionable environmental impacts.

6. CREDIT AGE

New Credits: Recently issued credits.
Vintage Credits: Older credits that have been issued for some time.

Reference link: https://gemini.google.com/ & https://www.lythouse.com/wp-content/uploads/2024/08/carbon-credits-1.webp

VERIFIED CARBON STANDARD (VCS)

Verified Carbon Standard (VCS) Program drives finance toward activities that reduce and remove emissions, improve livelihoods, and protect nature. VCS projects have reduced or removed more than one billion tons of carbon and other GHG emissions from the atmosphere. The VCS Program is a critical and evolving component in the ongoing effort to protect our shared environment. By marrying scientific rigor and transparency with innovative thinking, the VCS Program has continually brought new projects, organizations, and people into the voluntary carbon market, as well as a growing number of compliance markets, and given them the necessary confidence to participate.

WHY THE VCS?

The positive outcomes of carbon projects are underpinned by Verra's diligence and transparency: buyers of credits issued by Verra's VCS Program feel confident that their purchase had its intended effect; carbon credit sellers want markets with quality control guarantees, which help attract a solid pool of buyers.

The VCS Program has become the largest GHG crediting program in the world because of its rigorous rules and requirements; its adaptiveness to new scientific, technological, and regulatory developments; and the transparent information about its projects and their activities, which is publicly available on the Verra Registry.

These characteristics result in GHG emission reductions or removals that are real, measurable, additional, permanent, independently verified, conservatively estimated, uniquely numbered, and transparently listed (VCS Quality Assurance Principles).

The VCS Program has been endorsed by the International Carbon Reduction and Offset Alliance (ICROA) and meets the ICROA Code of Best Practice (external).

Source: https://www.controlunion.com/wp-content/uploads/2023/06/verified_carbon_standard_jpg.png
Reference link : https://verra.org/programs/verified-carbon-standard/

INTEGRITY COUNCIL FOR VOLUNTARY CARBON MARKET (ICVCM)

Integrity Council for Voluntary Carbon Market (ICVCM)

The Integrity Council for the Voluntary Carbon Market is a non-profit, independent governance body that aims to set and maintain a global standard for high integrity in the voluntary carbon market, unlocking private climate and carbon finance that would not otherwise be deployed. It was set up in 2021 in response to the final recommendations of the Taskforce on Scaling the Voluntary Carbon Markets (TSVCM), an initiative backed by more than 250 organisations.

AIMS

We are assessing carbon-crediting programs and methodology types against the Core Carbon Principles (CCPs). CCP labelled carbon credits will bring integrity to the market.

We are engaging with all stakeholders in the market including, VCM practitioners, governments and regulators, Indigenous Peoples & local communities.

. We are working to ensure carbon programs and projects increase ambition over time. Our stakeholder workshops will identify best practice in key areas to feed into the next iteration of the CCPs.

Source: https://www.ecosystemmarketplace.com/wp-content/uploads/2022/03/ICVCM-Logo.png
Reference link: https://icvcm.org/about-us/#

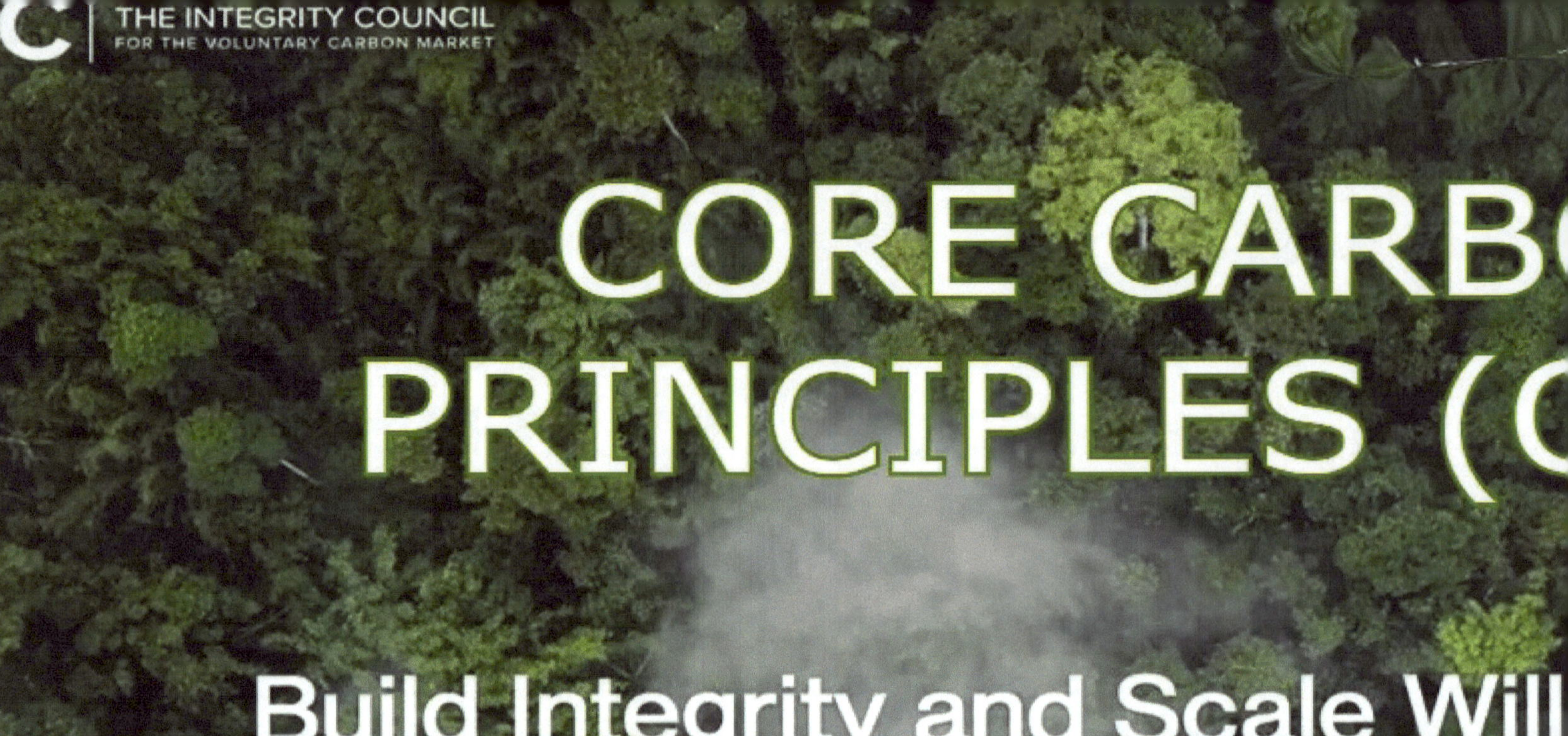

CORE CARBON PRINCIPLE (CCP)

The Core Carbon Principles (CCPs) are ten fundamental, science-based principles for identifying high-quality carbon credits that create real, verifiable climate impact. Developed with input from hundreds of organisations, they set a global benchmark for high integrity in the voluntary carbon market to raise it to a consistent level of quality and ensure it accelerates progress towards the 1.5°C target.

Governance

1. Effective governance
The carbon-crediting program shall have effective program governance to ensure transparency, accountability, continuous improvement and the overall quality of carbon credits.

2. Tracking
The carbon-crediting program shall operate or make use of a registry to uniquely identify, record and track mitigation activities and carbon credits issued to ensure credits can be identified securely and unambiguously.

3. Transparency
The carbon-crediting program shall provide comprehensive and transparent information on all credited mitigation activities. The information shall be publicly available in electronic format and shall be accessible to non-specialised audiences, to enable scrutiny of mitigation activities.

4. Robust independent third-party validation and verification
The carbon-crediting program shall have program-level requirements for robust independent third-party validation and verification of mitigation activities.

Emissions Impact

5. Additionality
The greenhouse gas (GHG) emission reductions or removals from the mitigation activity shall be additional, i.e., they would not have occurred in the absence of the incentive created by carbon credit revenues.

6. Permanence
The GHG emission reductions or removals from the mitigation activity shall be permanent or, where there is a risk of reversal, there shall be measures in place to address those risks and compensate reversals.

7. Robust quantification of emission reductions and removals
The GHG emission reductions or removals from the mitigation activity shall be robustly quantified, based on conservative approaches, completeness and scientific methods.

8. No double-counting
The GHG emission reductions or removals from the mitigation activity shall not be double counted, i.e., they shall only be counted once towards achieving mitigation targets or goals. Double counting covers double issuance, double claiming, and double use.

Sustainable Impact

9. Sustainable development benefits and safeguards
The carbon-crediting program shall have clear guidance, tools and compliance procedures to ensure mitigation activities conform with or go beyond widely established industry best practices on social and environmental safeguards while delivering positive sustainable development impacts.

10. Contribution toward net zero transition
The mitigation activity shall avoid locking-in levels of GHG emissions, technologies or carbon-intensive practices that are incompatible with the objective of achieving net zero GHG emissions by mid-century.

Source: https://carboncredits.b-cdn.net/wp-content/uploads/2022/07/core-carbon-principles-ICVCM.png
Reference link: https://icvcm.org/core-carbon-principles/#:~:text=The%20Core%20Carbon%20Principles%20%28CCPs%29%20are%20a%20global,set%20rigorous%20thresholds%20on%20disclosure%20-and%20sustainable%20development

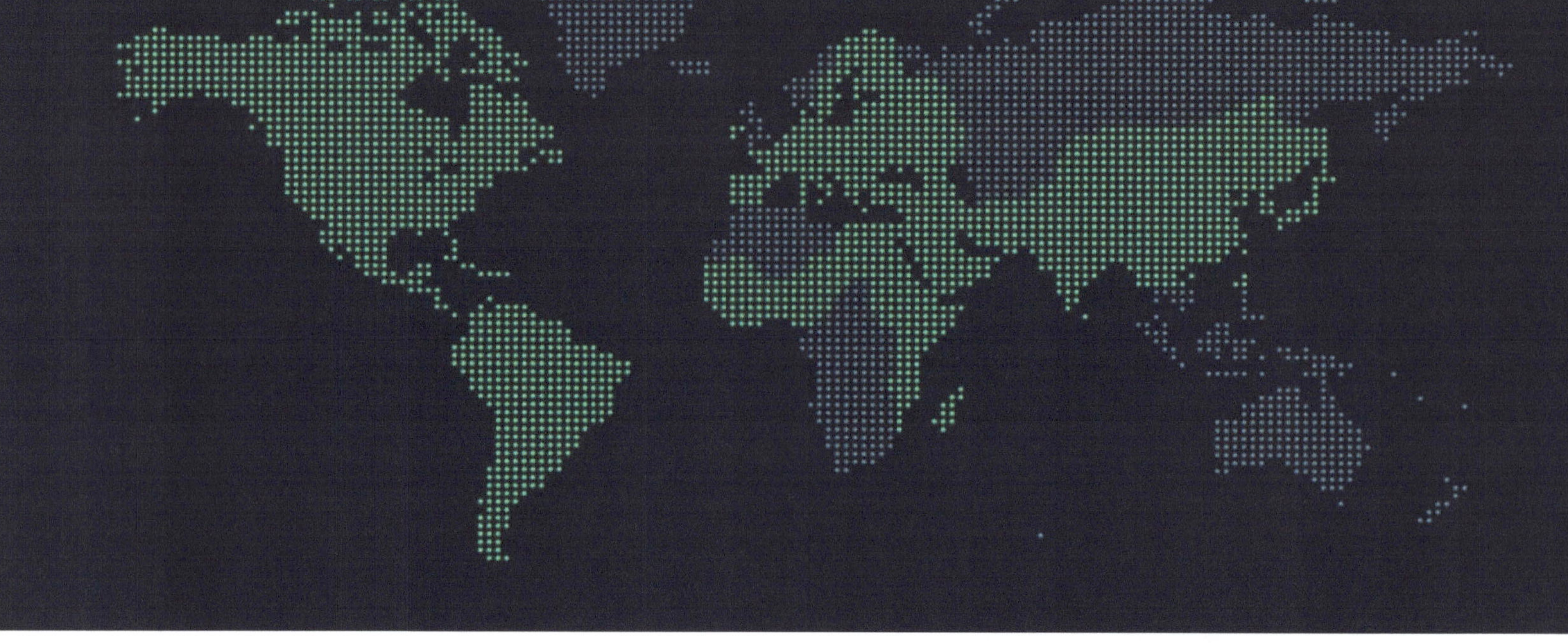

CLIMATE ACTION DATA TRUST(CADT)

Climate Action Data Trust (CAD Trust) is a decentralised metadata platform that links, aggregates and harmonises all major carbon registry data to enhance transparent accounting in line with Article 6 of the Paris Agreement. The CAD Trust open-source metadata system uses blockchain technology to create a decentralised record of carbon market activity with the aim to avoid double counting, increase trust in carbon credit data and build confidence in carbon markets through improved transparency.

How it works

CAD Trust Unifies Carbon Registry Data on a Single Platform Using a Common Data Model. CAD Trust is an open-source metadata system designed to share information about carbon credits and projects across digital platforms, facilitating the future integration of multiple registry systems. It is currently in the process of connecting various carbon credit registries. The goal is to offer free public access to this information in a harmonised and user-friendly format, enabling private companies, NGOs, and governments to utilise it for benchmarking, double counting risk checking, and compliance reporting.

Vision

Today's carbon markets are complex and fragmented. A prevalence of independent private registries, and national-level carbon markets, combined with a lack of centralised registries between voluntary and compliance markets poses a serious challenge to ensuring maximum transparency in the marketTo future-proof the market, CAD Trust offers a decentralised, blockchain-powered digital infrastructure that connects these registries and provides public access to the information in a harmonised and user-friendly format. Our objective is to maximise the transparency of carbon credits, minimise the risk of double counting, and enhance the overall integrity of the markets.With a decentralised and secure digital infrastructure, the services built by the public and private sector such as compliance reporting, transacting, and benchmarking can be easier and more effective.

Mission

We believe that transparency and accountability are vital for efficient carbon markets that drive climate action and the associated finance in public and private sectors.Climate Action Data Trust acts in the public interest to increase transparency, safeguard against double-counting and build further confidence in the market. We leverage carbon credit data from both compliance and voluntary markets, fostering collaboration, facilitating innovation, and helping unlock the potential of carbon markets across the world.

Reference link: https://climateactiondata.org/
Source: https://carboncredits.b-cdn.net/wp-content/uploads/2022/12/world-bank-CAD-trust-simulation-participants.jpg

GOLD STANDARD
(VOLUNTARY CARBON OFFSET PROGRAM)

Gold Standard for the Global Goals is an innovative impact standard bringing high-level integrity and credibility to the most important global goal of our time: climate security and sustainable development. Gold Standard is a not-for-profit headquartered in Geneva, Switzerland, focused on catalysing more ambitious climate action to achieve the global goals through robust standards and verified impacts.

Comprised of numerous methodologies, guidelines and requirements - backed by strong safeguards and principles – it enables the credible measurement, reporting and verification of positive impacts generated by climate and sustainable development initiatives. Its robust yet customisable design allows for application across a diverse range of initiatives and users, with one fundamental objective: driving financial support to climate and sustainable development initiatives.

Vision
Climate security and sustainable development for all.

Mission
To catalyse ambitious climate action to achieve the Global Goals through robust standards and verified impacts.

Gold Standard Principles
To ensure projects and programmes adhere to the highest levels of social and environmental integrity, are successful in the long-term, and maximise the positive impacts, Gold Standard has five key principles that all projects need to comply with to achieve Gold Standard certification.

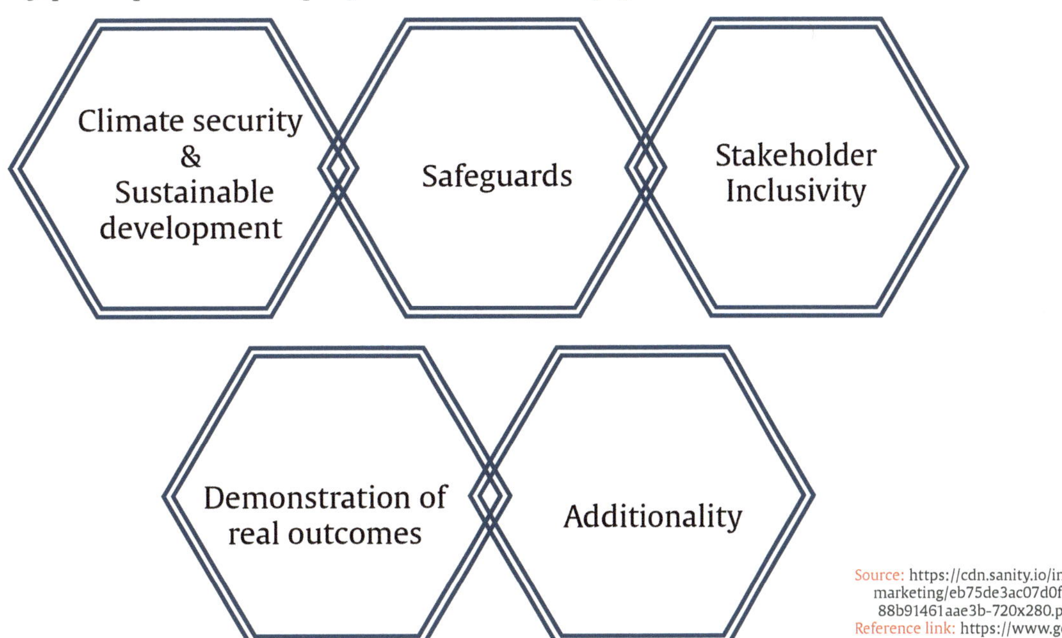

Source: https://cdn.sanity.io/images/f901zpue/
marketing/eb75de3ac07d0f702379e8336af-
88b91461aae3b-720x280.png?auto=format
Reference link: https://www.goldstandard.org/

CARBON OFFSETTING AND REDUCTION SCHEME FOR INTERNATIONAL AVIATION (CORSIA)

WHAT IS CORSIA?

CORSIA is the first global market-based measure for any sector and represents a cooperative approach that moves away from a "patchwork" of national or regional regulatory initiatives. It offers a harmonized way to reduce emissions from international aviation, minimizing market distortion, while respecting the special circumstances and respective capabilities of ICAO Member States. CORSIA complements the other elements of the basket of measures by offsetting the amount of CO_2 emissions that cannot be reduced through the use of technological improvements, operational improvements, and sustainable aviation fuels with emissions units from the carbon market.

Source: https://cdn.sanity.io/images/f901zpue/marketing/eb75de3ac07d0f702379e8336af88b91461aae3b-720x280.png?auto=format
Reference link: https://www.goldstandard.org/

CORSIA SCOPE AND TIMELINE

● CORSIA operates on a route-based approach and applies to international flights, i.e. flights between two ICAO States. A route is covered by CORSIA offsetting requirements if both the State of departure and the State of destination are participating in the scheme, and is applicable to all aeroplane operators (i.e. regardless of the administering State) on the route.

● All aeroplane operators with international flights producing annual CO_2 emissions greater than 10,000 tonnes from aeroplanes with a maximum take-off mass greater than 5,700 kg, are required to monitor, verify and report their CO_2 emissions on an annual basis from 2019. The average CO_2 emissions reported during 2019 and 2020 represents the baseline for carbon neutral growth from 2020, and so the aviation sector is required to offset any international CO_2 emissions above that level.

● CORSIA includes three implementation phases. During the pilot and first phases, offsetting requirements will only be applicable to flights between States which have volunteered to participate in CORSIA offsetting. As of 1 January 2022, 107 States have volunteered representing an increase of 19 States from 2021. The second phase applies to all ICAO Contracting States, with certain exemptions.

● In June 2020, the ICAO Council decided that the baseline during the pilot phase of CORSIA (2021-2023) would be 2019 emissions only. This decision was made to safeguard the aviation industry against inappropriate economic burden resulting from an unexpected lower baseline caused by the COVID-19 pandemic that saw an approximate 56% decline in aviation emissions in 2020 compared to 2019. It is expected that, due to the change of baseline and the foreseen recovery scenario of international aviation after the COVID-19 pandemic, there will be no offsetting requirements for airlines during the pilot phase of the scheme.

EXCHANGE TRADED CARBON CREDIT CONTRACTS

EXCHANGE-TRADED CARBON CREDIT CONTRACTS?

These contracts allow participants to buy and sell carbon credits on regulated exchanges, similar to trading stocks or commodities. Each contract typically represents a specific quantity of carbon credits.

KEY FEATURES

Standardization: Contracts are standardized, making them easier to trade and more transparent in pricing.

Liquidity: Trading on an exchange usually provides higher liquidity compared to OTC markets, enabling faster buying and selling.

Regulation: Exchange-traded contracts are subject to oversight by regulatory bodies, which helps ensure fair practices and transparency.

Accessibility: Investors and companies can access these markets easily, broadening participation.

EXAMPLES OF EXCHANGES

Pansive CBL: Focuses on blockchain technology to enhance transparency and traceability in carbon credit trading.

European Energy Exchange (EEX): A major platform for trading EU Emission Allowances (EUAs).

Intercontinental Exchange (ICE): Offers a variety of carbon credit contracts, including compliance and voluntary credits.

ADVANTAGES

Market Efficiency: The structured environment promotes price discovery and reduces information asymmetry.

Reduced Counterparty Risk: Clearinghouses often guarantee trades, reducing the risk of default by participants.

Broader Participation: Attracts institutional investors alongside corporations seeking to manage their carbon footprints.

Reference link: http://www.chatgpt.com

OVER-THE-COUNTER (OTC) CARBON CREDIT CONTRACTS

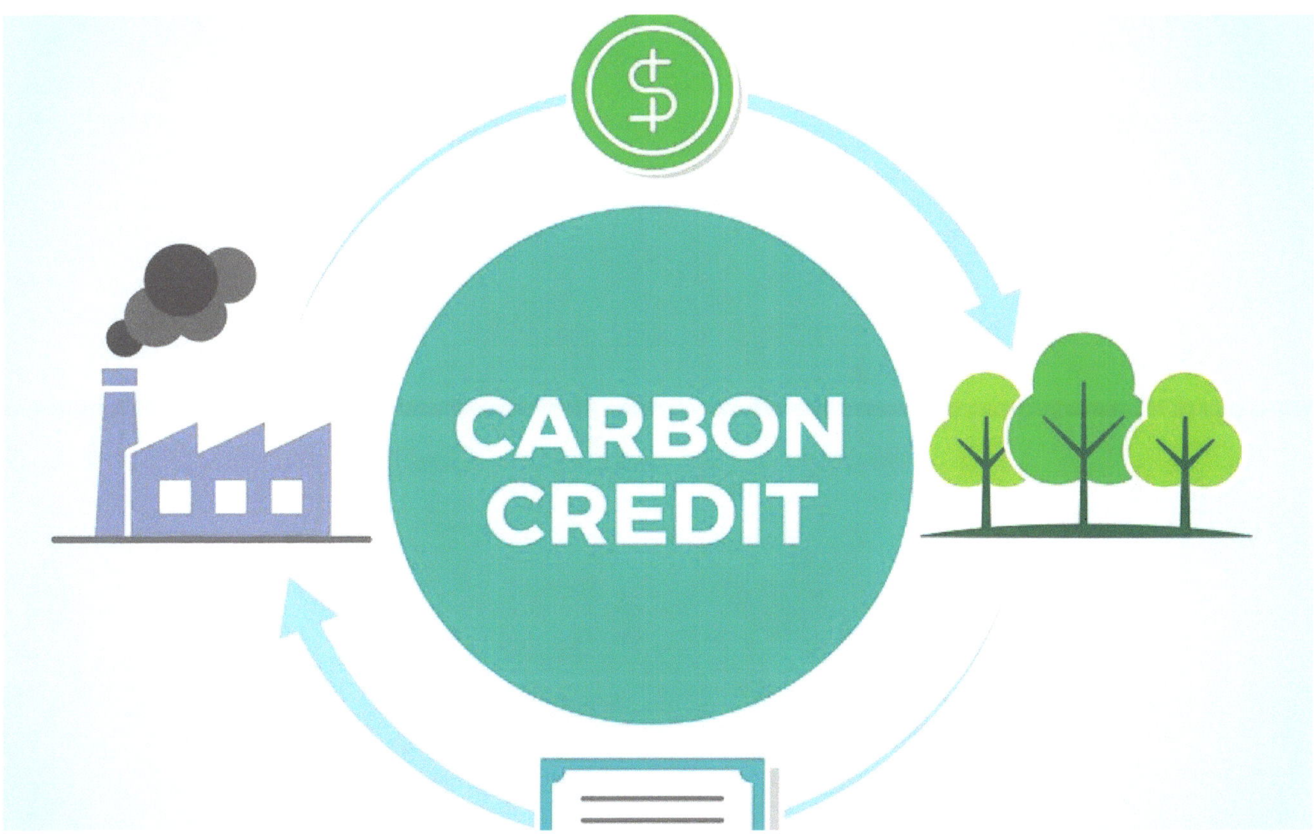

Carbon credits represent the right to emit one metric ton of carbon dioxide or an equivalent amount of other greenhouse gases. They are part of efforts to reduce global emissions and combat climate change

OTC MARKET

Direct Negotiation: OTC trades allow for flexibility and customized terms between parties.
Types of Contracts: These can include spot contracts (for immediate delivery) or futures contracts (for delivery at a later date).
Parties Involved: Participants typically include corporations, NGOs, and sometimes governments looking to offset emissions or invest in sustainability.

ADVANTAGES OF OTC CONTRACTS

Customization: Tailored agreements to suit specific needs.
Privacy: Transactions are not publicly disclosed, which can be advantageous for certain buyers or sellers.
Access: Smaller or specialized projects may find it easier to trade in the OTC market.

Reference link: http://www.chatgpt.com

REDD+ stands for "Reducing Emissions from Deforestation and Forest Degradation, plus conservation of forest carbon stocks, sustainable management of forests, and enhancement of forest carbon stocks."

Discussions first emerged in 2005 led by Papua New Guinea at a meeting of the Coalition of Rainforest Nations, held in Montreal during the 11th Conferences of the Parties (COP11) to the United Nations Framework Convention on Climate Change (UNFCCC). The idea was to create a potential results-based finance support mechanism aimed at reducing deforestation.

This led to the introduction in 2013 of the Warsaw REDD+ framework as a mechanism enabling nations to financially support the efforts of other countries in reducing deforestation. Notably, this framework did not permit the generation of carbon credits.

Alongside REDD+ negotiations under the UNFCCC, various public schemes such as the United Nations Collaborative Programme on Reducing Emissions from Deforestation and Forest Degradation in Developing Countries (UN-REDD) and the Forest Carbon Partnership Facility (FCPF) of the World Bank were developed, while private actors, such as the Voluntary Carbon Standard (VCS) also emerged.

Private Standards like the VCS, a popular voluntary scheme used by private organisations, generate carbon credits from forestry projects. These are often used for "offsetting" purposes. Generating and selling tradable carbon credits is significantly different from the initial concept of financing ecosystem services under the UNFCCC, specifically because purchasing carbon credits enables buyers to meet emission reduction targets, whereas under the UNFCCC mechanism this is not allowed.

Reference link: https://carbonmarketwatch.org/2023/11/09/redd-faq/

ECOSYSTEM MARKETPLACE

Ecosystem Marketplace, an initiative of Forest Trends, is a non-profit organization based in Washington, DC, that focuses on increasing transparency and providing information for ecosystem services and payment schemes.

The idea of launching Ecosystem Marketplace was borne out of meeting by members of The Katoomba Group, an international working group composed of leading experts from forest and energy industries, research institutions, the financial world, and environmental NGOs dedicated to advancing markets for some of the ecosystem services provided by forests – such as watershed protection, biodiversity habitat, and carbon capture and storage. Both Ecosystem Marketplace and The Katoomba Group are initiatives under Forest Trends. Ecosystem Marketplace specializes in market-based approaches to environmental protection—in the public and private spheres—regarding greenhouse gases, water, biodiversity, and conservation. This is why it is perceived as an organisation that pushes for neoliberal biodiversity conservation.

Ecosystem Marketplace's website states: "We believe that by providing solid and trust-worthy information on prices, regulation, science, and other market-relevant issues, markets for ecosystem services will one day become a fundamental part of our economic and environmental system, helping give value to environmental services that have, for too long, been taken for granted

MARKET COVERAGE

Ecosystem Marketplace has covered information on a number of areas, including the following:

Carbon

Relevant material can be accessed via the main Ecosystem Marketplace website, the carbon markets landing page, and the Forest Carbon Portal.

- Kyoto-related carbon markets (EU ETS, CDM, JI)
- Voluntary over-the-counter markets
- North American state and regional markets
- Payments for terrestrial carbon sequestration

Water

Relevant material can be accessed via the main Ecosystem Marketplace website and Watershed Connect.
- Water quality and nutrient trading
- Schemes to pay for watershed services
- Nitrogen offset programs

Biodiversity

Relevant material can be accessed via the main Ecosystem Marketplace website and the Species Banking Portal.
- Conservation banking in the U.S.
- Wetland/stream mitigation banking in the U.S.
- Australian biodiversity offset programs
- Voluntary biodiversity offsets

Communities

Relevant material can be accessed via the main Ecosystem Marketplace website and the Communities Portal

- PES program design
- community ecosystem management
- community buyer relations
- poverty equity
- community as buyer
- payments compensation
- land tenure rights

Source: https://www.ecosystemmarketplace.com/
Reference link: https://en.wikipedia.org/wiki/Ecosystem_Marketplace

IETA was the first international, multi-sectoral, purely business group devoted to pricing and trading greenhouse gas reductions. From the start, it had a strong focus on the Kyoto mechanisms and helped members using, hosting and investing in Clean Development Mechanism (CDM) and Joint Implementation (JI) projects by disseminating policy and market information and promoting development and reform. As markets developed, IETA expanded its work to cover the EU ETS, sub-national efforts in North America, China, Korea, New Zealand and emerging markets across Latin America. In more recent years, IETA has also developed work streams on digitisation, aviation, natural climate solutions, voluntary carbon markets, and carbon removals

IETA's mission is to:

• Empower business to engage in climate action and pursue net zero ambitions to advance the Paris Agreement's objectives, and

• Establish effective market-based trading systems for GHG emissions and reductions that are environmentally robust, fair, open, efficient, accountable and consistent across national boundaries.

In pursuit of our mission, IETA works in collaboration with other stakeholders to:

Develop components of the GHG market and trading systems At the international level, IETA focuses on the transparency rules and accounting standards under Articles 5 and 6 of the Paris Agreement, and also for the Carbon Offset and Reduction Scheme for International Aviation (CORSIA) and voluntary markets. At the national and sub-national levels, IETA promotes best practices for offset and allowance markets aligned with the goal of reaching net-zero emissions by 2050.

Develop a global GHG market:

IETA is dedicated to the establishment of linked trading systems to ensure efficient and competitive GHG markets. IETA provides thought leadership on the distinct advantages of linked markets by collaborating with deep subject matter experts in academia and think-tanks.

Strengthen business capacity and embrace innovation:

IETA brings together experienced carbon market practitioners and stakeholders in new emissions markets to share lessons learned and best practices. IETA seeks continuous improvement in carbon pricing systems, including innovations to advance natural climate solutions.

Promote market-based solutions and broad participation in GHG markets:

IETA uses its global reach and reputation to promote the go-to events for market participants, including its many regional carbon fora. These events showcase excellence, explore market insights and offer networking opportunities in every region of the world.

SCIENCE BASED TARGETS
INITIATIVE (SBTI)

The Science Based Targets initiative (SBTi) is a corporate climate action organization that enables companies and financial institutions worldwide to play their part in combating the climate crisis. We develop standards, tools and guidance which allow companies to set greenhouse gas (GHG) emissions reductions targets in line with what is needed to keep global heating below catastrophic levels and reach net-zero by 2050 at latest. The SBTi is incorporated as a charity, with a subsidiary which will host our target validation services. Our partners are CDP, the United Nations Global Compact, the We Mean Business Coalition, the World Resources Institute (WRI), and the World Wide Fund for Nature (WWF).

Vision

By 2050, the world will have transitioned towards a net-zero and equitable economy that serves the needs of the population within the limits of the planet.

Mission

To drive science-based climate action in the corporate sector consistent with limiting warming to 1.5°C.

AIMS

The SBTi mobilizes the private sector to take the lead on urgent climate action.
The SBTi enables companies and financial institutions to understand how much and how quickly they must decarbonize to prevent the worst impacts of climate change.
We enable them to tackle global warming.
Action is needed in key sectors, many of which require tailored approaches to setting targets. The SBTi provides clarity and guidance on these journeys, including for the forest, land and agriculture and finance sectors.
In 2021 the SBTi launched the world's first Corporate Net-Zero Standard, to ensure that companies' net-zero targets are consistent with what is required to achieve net-zero no later than 2050.

Source: https://www.southpole.com/uploads/thumbs/sbti-validates-south-poles-2040-science-based-net-zero-target-_1920x768.jpg
Reference link: https://sciencebasedtargets.org/about-us

VOLUNTARY CARBON MARKETS INTEGRITY INITIATIVE (VCMI)

VCMI MISSION

The Voluntary Carbon Markets Integrity Initiative (VCMI) enables high-integrity voluntary carbon markets which contribute to the goal of the Paris Agreement, bringing benefits for people and the planet. The Voluntary Carbon Markets Integrity Initiative (VCMI) is an international non-profit organization with a mission to enable high-integrity voluntary carbon markets (VCMs) that deliver real and additional benefits to the atmosphere, help protect nature, and accelerate the transition to ambitious, economy-wide climate policies and regulation.

The organization is fully aligned with the goals of the Paris Agreement and is committed to a world on track to 1.5 degrees and net zero emissions by mid-century, achieved through a just transition that enhances equality and sustainable development for all.

WHAT WE DO

- •VCMI collaborates with stakeholders from civil society, the private sector, Indigenous Peoples, local communities, and governments, to realize the full potential of high-integrity voluntary carbon markets.

- •It works to foster a shared vision for high-integrity VCMs to make a meaningful contribution to climate action while also supporting the achievement of the UN SDGs. VCMI connects and aligns with, and amplifies, initiatives that share this vision.

- •On the demand-side, the VCMI Claims Code of Practice is a rulebook on how companies can make voluntary use of carbon credits as part of credible, science-aligned net-zero decarbonization pathways. It builds trust and confidence in how companies engage with VCMs, assisting them in making credible climate claims.

- •On the supply-side, the VCM Access Strategy Toolkit provides guidance for countries to engage in high-integrity VCMs in support of national climate and economic prosperity.

Reference link: https://vcmintegrity.org/about/
Source: https://vcmintegrity.org/wp-content/uploads/2021/07/about-seeds-1.jpg

EMISSION REDUCTION – LINKED BOND OF WORLD BANK

HOW DOES THE BOND WORK?

The Emission Reduction-Linked Bond is a USD 50 million 5-year principal protected bond issued by the International Bank for Reconstruction and Development ("IBRD" or the "World Bank") with a unique feature, whereby investors return is linked to the issuance of Verified Carbon Units ("VCUs") from a water purification project in Vietnam (the "Project").

With the bond, investors do not receive ordinary coupon payments that are usually paid by regular bonds. Instead, an amount equal to the cash flows that would have been paid as coupons on a regular bond are "frontloaded" and - through a hedge transaction with Citi - used to support the financing of the Project. The Project aims to improve the quality of drinking water in schools and institutions throughout Vietnam in a climate friendly manner, and in doing so to generate and issue VCUs on the Verra Registry. The bond investors will receive variable coupon payments based on the number of VCUs issued from the Project from time to time.

Will the bond be listed?

The Bond will be listed on the Luxembourg Stock Exchange.

Will the bond be rated?

The bond is expected to be rated AAAp by Standard & Poor's.

Who is the issuer of the bond?
The bond is issued by the International Bank for Reconstruction and Development ("IBRD" or the "World Bank").

Will the bond give carbon credits to investors?

No. The bond will pay investors a VCU-Linked Interest Amount from time to time based on the number of VCUs generated by the Project.

How will the bond proceeds be used?

The bond issuance proceeds will be used for the World Bank's sustainable development activities. World Bank bonds support the financing of a combination of green and social, i.e., "sustainable development", projects, programs, and activities in IBRD member countries. See IBRD's Sustainable Development Bond Framework for examples of eligible projects, programs, and activities. The water purifier Project is not a World Bank project

CARBON CREDIT LINKED BOND

Definition:

CCLBs are debt securities whose returns are linked to the performance of carbon credits generated by a specific project or portfolio of projects.

Key Features:

- Returns tied to carbon credit prices or volumes
- Proceeds fund projects generating carbon credits (e.g., renewable energy, energy efficiency)
- Investors benefit from growing demand for carbon credits
- Bonds can be issued by governments, corporations, or financial institutions

BENEFITS:

- Mobilizes funding for low-carbon projects
- Provides investors with exposure to carbon markets
- Supports global climate goals by promoting emissions reductions

STRUCTURE:

Issuer	Underlying Assets
Entity issuing the bond (e.g., government, company)	Carbon credits generated by the funded project(s)
Return Mechanism	**Credit Rating**
Returns linked to carbon credit prices or volumes	Bonds may be rated by credit rating agencies

Examples:

The World Bank's Green Bond program has issued CCLBs to support renewable energy projects. Corporations like Enel and Engine have issued CCLBs to fund low-carbon initiatives.

INTERNATIONALLY TRANSFERRED MITIGATION OUTCOMES (ITMO)

Article 6 of the Paris Agreement allows for the use of co-operative approaches, such as internationally transferred mitigation outcomes (ITMOs), to increase ambition and promote sustainable development. However, rules involving the use of ITMOs have yet to be agreed-upon by the parties to the United Nations Framework Convention on Climate Change (UNFCCC). Currently, no countries are on track to meet their nationally determined contributions (NDCs). Climate ambition needs to be significantly scaled up to limit average global temperature rise to 1.5°C to avoid planetary tipping points caused by climate change. The final negotiated text for Article 6 will define our ability to achieve this goal.

● After 20 years of experience with the Clean Development Mechanism (CDM) and Joint Implementation (JI), several accounting and credibility concerns have been identified. These concerns must be addressed, and lessons learned must be applied in the design of ITMOs.

● ITMOs must adhere to the basic rules of a credible offset mechanism. These include ensuring credited emissions reductions are additional to what would have occurred in the absence of the ITMO, can be verified as reductions from a credibly determined and agreed-upon benchmark, and are monitored and reported on over time.

● ITMOs represent a fundamental change from CDM and JI in two important ways. First, all countries, not just developed countries, have now set emissions targets under their NDCs. Within this new context, ITMOs should be used in accordance with the objectives of the Paris Agreement, to increase global climate ambition and not create perverse incentives that impair ambition. Second, while the types of activities and projects eligible to generate mitigation outcomes under Article 6 remain a matter of debate, the eligibility of low-carbon exports (not eligible under CDM/JI), in particular, raises significant challenges. These concerns have yet to be fully addressed and are critical to the integrity and effectiveness of international co-operation on climate action.

FOUR PRINCIPLES OF CREDIBLE ITMOS

The following four principles should be used to evaluate whether an ITMO is credible. Only by adhering to these principles should ITMOs be supported as a credible component of an international emissions mitigation strategy.

●Raises global ambition: Emissions credits eligible for ITMOs must be for reductions that go above and beyond the ambition of current NDC targets, not used as a pathway to achieve current NDC targets.

●Achieves additional emissions reductions: It must be demonstrable that the ITMO achieves emissions reductions that would not otherwise have occurred in the absence of the ITMO, and that the emissions reduced by the ITMO have not been displaced or leaked to another location.

●Achieves verifiable emissions reductions: It must be verifiable that the ITMO achieves emissions reductions against a globally agreed-upon baseline and can be monitored and reported on over time under clear monitoring guidelines.

●Protects human rights and promotes the health and wellbeing of all communities: ITMOs must respect the rights of all communities and Indigenous Peoples by avoiding undue harm, thoroughly assessing risks, supporting reconciliation, and including safeguards for human rights protection.

Source: https://www.cigionline.org/static/images/iStock-612252326.2e16d0ba.fill-1600x900.jpg & https://www.rainforestcoalition.org/wp-content/uploads/2023/07/CfRN_JULY_AUGUST_2023_ITMO-event_article-scaled.jpg Reference link: https://www.pembina.org/reports/itmos-backgrounder-2019.pdf

LOW CARBON ECONOMY

A low-carbon economy (LCE) is an economy which absorbs as much greenhouse gas as it emits. Greenhouse gas (GHG) emissions due to human activity are the dominant cause of observed climate change since the mid-20th century. There are many proven approaches for moving to a low-carbon economy, such as encouraging renewable energy transition, energy conservation, electrification of transportation (e.g. electric vehicles), and carbon capture and storage. An example is zero-carbon cities.

RATIONALE AND AIMS

GHG emissions due to human activity are the dominant cause of observed climate change since the mid-20th century. Continued emission of greenhouse gases will cause long-lasting changes around the world, increasing the likelihood of severe, pervasive, and irreversible effects for people and ecosystems.

Nations may seek to become low-carbon or decarbonized economies as a part of a national climate change mitigation strategy. A comprehensive strategy to mitigate climate change is through carbon neutrality.

METHODS

Achieving a low-carbon economy involves reducing greenhouse gas emissions in all sectors that produce greenhouse gases, for example energy, transportation, industry, and agriculture. The literature often speaks of a transition from a high-carbon economy to a low-carbon economy. This transition should take place in a just manner (this is termed just transition). There are many strategies and approaches for moving to a low-carbon economy, such as encouraging renewable energy transition, efficient energy use, energy conservation, electrification of transportation (e.g. electric vehicles), carbon capture and storage, climate-smart agriculture. This requires for example suitable energy policies, financial incentives (e.g. emissions trading, carbon tax), individual action on climate change, business action on climate change.

Source: https://www.esi-africa.com/wp-content/uploads/2017/04/Green-Energy-Copyright-a-hrefhttpswww.123rf.comprofile_lassedesignenlassedesignen-123RF-Stock-Photoa.jpg
Reference link: https://en.m.wikipedia.org/wiki/Low-carbon_economy#-ww

CARBON TRADE EXCHANGE (CTX)

Carbon Trade Exchange (CTX) operates spot exchanges in multiple global environmental commodity markets, including Carbon, Renewable Energy Certificates (RECs) and Water. CTX is currently operating in Australia, the EU and the US, with the ability to expand into other jurisdictions and markets.

The commercialized technology platforms that we operate allow us to cover the various products that exist within multiple markets and provide an exchange by which sellers and buyers are matched in a secure and efficient manner. The exchange provides a trusted global marketplace which removes risk in the delivery-vs-payment process, making it easier for businesses of all sizes to invest in sustainable, clean-tech, and energy-efficient projects around the world in support of a more efficient and low-carbon economy.

The CTX exchange platform uniquely interfaces with multiple environmental commodity registries, including APX, Climate Action Reserve, American Carbon Registry, VCS, Gold Standard and the Australian REC Registry. It also electronically links to financial intermediaries, such as Westpac in Australia, to provide efficient trading and create liquid and transparent markets.

CARBON NEUTRAL CERTIFICATION

Carbon neutral certification is a verification process that confirms a company, organization, or product has achieved net-zero greenhouse gas (GHG) emissions. This means that the total amount of carbon dioxide and other greenhouse gases emitted is balanced by an equivalent amount removed from the atmosphere.

HOW DOES CARBON NEUTRAL CERTIFICATION WORK?

Measurement: The organization calculates its carbon footprint, including direct emissions (Scope 1), indirect emissions from purchased energy (Scope 2), and indirect emissions from the value chain (Scope 3).

Reduction: The organization implements strategies to reduce its carbon emissions, such as improving energy efficiency, switching to renewable energy sources, and optimizing transportation.

Offsetting: Any remaining emissions are offset by investing in projects that capture or remove carbon dioxide from the atmosphere, such as planting trees or supporting renewable energy initiatives.

BENEFITS OF CARBON NEUTRAL CERTIFICATION

Environmental Impact: Contributes to mitigating climate change and reducing the negative effects of greenhouse gas emissions.

Brand Reputation: Enhances the organization's image as environmentally responsible and socially conscious.

Customer Loyalty: Attracts environmentally conscious consumers who are more likely to support businesses with sustainable practices.

Competitive Advantage: Differentiates the organization from competitors and positions it as a leader in sustainable business practices

Reference link: https://gemini.google.com/ & https://auramag.in/wp-content/uploads/2022/05/Climate-Change-Threat-or-Opportunity-@aura-emagazine.jpg Source: https://www.controlunion.com/wp-content/uploads/2023/06/cu_carbon_neutral_logo_23-11.jpg

NATIONAL ACTION PLAN ON CLIMATE CHANGE

National Action Plan for Climate Change (NAPCC) is a Government of India's programmed launched in 2008 to mitigate and adapt to the adverse impact of climate change. The action plan is designed and published under the guidance of Prime Minister's Council on Climate Change (PMCCC).

The 8 sub-missions aimed at fulfilling India's developmental objectives with focus on reducing emission intensity of its economy. The plan will rely on the support from the developed countries with the prime focus of keeping its carbon emissions below the developed economies at any point of time. The 8 missions under NAPCC are as follows:

- National Solar Mission
- National Mission for Enhanced Energy Efficiency
- National Mission on Sustainable Habitat
- National Water Mission
- National Mission for Sustaining Himalayan Ecosystem
- Green India Mission
- National Mission for Sustainable Agriculture
- National Mission on Strategic Knowledge for Climate Change

CARBON CREDIT TRADING SCHEME

Carbon credit trading system in India assigns a value, known as a carbon credit, to every ton of carbon dioxide equivalent (tCO_2e) reduced or avoided, providing a structured framework for the country's carbon market. Industries will play a major role in contributing to India's emission reduction goals under this scheme. The Indian government is working on its system of accredited carbon verifiers and a fully functional carbon credit trading system (CCTS) to help Indian exporters of steel and aluminum meet the prescribed EU carbon norms by January 2026. Carbon credit in India is traded on NCDEX only as a future contract.

Obligated Entities

A Carbon Credits Trading System in Obligated Entities is a regulatory framework that requires entities (such as industries or companies) to offset a portion of their greenhouse gas emissions by purchasing carbon credits. These credits are generated by projects that reduce emissions or remove CO_2 from the atmosphere.

Examples:

European Union's Emissions Trading System (EU ETS): A cap-and-trade system for power, industry, and aviation sectors. California's Cap-and-Trade Program: A mandatory carbon credits trading system for power, industry, and transportation sectors. India's Perform, Achieve, and Trade (PAT) scheme: A mandatory carbon credits trading system for energy-intensive industries like cement, steel, and power.

Non-Obligated Entities

A voluntary market-based system that allows entities not required to reduce greenhouse gas emissions (Non-Obligated Entities) to buy, sell, and trade carbon credits to offset their emissions or invest in low-carbon projects, promoting sustainable development and emissions reductions.

Decarbonization Projects

The primary objective of developing the Indian Carbon Market is to support the decarbonization of the Indian economy. By pricing greenhouse gas (GHG) emissions through the trading of Carbon Credit Certificates, the ICM aims to incentivize industries and entities to adopt low-carbon pathways. Decarbonization Projects are initiatives that aim to reduce greenhouse gas emissions and carbon footprint, transitioning from fossil fuels to cleaner energy sources, and promoting sustainable development.

Example:

Renewable Energy Projects: Solar farms, wind farms, hydroelectric power plants, geothermal energy projects. Energy Efficiency Projects: Building insulation, LED lighting, smart grids, industrial process optimization.

Offset Mechanism

Offset mechanisms are strategies used to compensate for unavoidable greenhouse gas emissions by reducing or removing equivalent amounts of emissions elsewhere. These mechanisms allow individuals, organizations, or countries to offset their emissions by investing in projects or activities that reduce greenhouse gas emissions.

Example

A company in the US wants to offset the emissions from its air travel, so it buys credits from a wind farm project in Mexico that reduces emissions by 1,000 tons of CO_2 per year. The company pays for the credits, which helps finance the wind farm, and can then claim the equivalent emissions reduction towards its own sustainability goals, effectively offsetting the impact of its air travel.

Validation

CCTS (Carbon Credit Trading System) validation is the process of verifying the quality and authenticity of carbon credits generated from projects that reduce greenhouse gas emissions. It includes

1. Project Registration: The project developer registers their project with the CCTS, providing detailed information about the project's design, location, and emissions reduction potential.
2. Documentation Review: The CCTS reviews the project's documentation to ensure it meets the required standards and criteria.

Example

A wind farm project is registered with the Carbon Credit Trading System (CCTS) to generate carbon credits. The project reduces 50,000 tons of CO_2 emissions per year by generating 100,000 MWH of renewable energy. The CCTS reviews the project's documents, conducts a site visit, and verifies the emissions reduction calculations to ensure accuracy. After validation, the CCTS issues a report approving the project, allowing it to generate 50,000 carbon credits annually.

Verification

Verification is the process of confirming that a project's emissions reductions have occurred as claimed. This involves:

1. Monitoring: Tracking the project's emissions reductions through data collection and monitoring systems.
2. Reporting: Submitting regular reports to the CCTS, detailing the project's emissions reductions.
3. Audit: Conducting annual audits to ensure the reported emissions reductions are accurate and reliable.

Certification:

Certification is the issuance of a certificate by the CCTS, confirming that a project's emissions reductions have been verified and meet the required standards. This certificate is known as a Verified Emission Reduction (VER) or Certified Emission Reduction (CER). The certificate includes:

1. *Project details:* Project name, location, and type.
2. *Emissions reductions:* Verified number of emissions reduced.
3. *Verification period:* Timeframe for which emissions reductions were verified.
4. *Certificate serial number:* Unique identifier for the certificate.

Reference link: http://www.chatgpt.com

CARBON CREDIT CERTIFICATE (CCC)

The carbon credit certification process plays a crucial role in the fight against climate change, as businesses and organizations strive to decrease their environmental impact. Carbon credits are tradable certificates that represent the reduction or removal of one metric ton of carbon dioxide equivalent (CO_2e) emissions.

They are used by companies to offset their own emissions or by governments to meet emissions reduction targets. There are several certification bodies and standards that ensure the environmental integrity of these carbon credits, providing a clear and transparent framework for their issuance and verification.

To obtain carbon credit certification, project developers must submit their projects for rigorous assessment and verification by independent certification bodies. These bodies evaluate the project's emission reduction methodologies, its additionality (the extent to which the project results in emissions reductions beyond what would have occurred in the absence of the project), and the permanence of the emissions reductions. Once certified, the carbon credits are issued and can be bought or sold in the voluntary carbon market, which helps encourage more investments in emissions reduction projects.

- Carbon credit certification is essential for ensuring the environmental integrity of emission reduction efforts.
- Independent certification bodies assess and verify projects' methodologies, additionality, and the permanence of emissions reductions.
- The voluntary carbon market is driven by the demand for certified carbon credits, fostering further investment in emissions reduction projects.

CONVERTED CCCS (C-CCC)

Converted CCCs (C-CCC) is a type of carbon credit that arises from the conversion of existing emission reduction projects into carbon credit projects.

HOW IT WORKS:

Existing Project:
There's an existing project that has been reducing emissions but hasn't been generating carbon credits. For example, a reforestation project might have been implemented to sequester carbon, but it hasn't been registered under a carbon credit scheme.

Conversion:
This project is assessed to determine its potential to generate carbon credits. If it meets the criteria, it can be converted into a carbon credit project.

Credit Generation:
Carbon credits are issued based on the verified emissions reductions achieved by the project.

Benefits of CCCs:

Leveraging Existing Efforts:
CCCs can recognize and reward existing emissions reduction projects that might not have been formally acknowledged.

Additional Revenue:
The generation of carbon credits can provide additional revenue to project developers, which can help sustain their efforts.

Increased Market Supply:
CCCs can increase the supply of carbon credits, potentially leading to lower prices and making carbon offsetting more affordable.

Challenges and Considerations:

Retroactivity: Converting existing projects can sometimes be challenging due to the need to gather historical data and verify emissions reductions.

Project Eligibility: Not all existing projects are eligible for conversion. The project must meet specific criteria, such as having verifiable emissions reductions and being aligned with recognized standards.

Market Acceptance: CCCs may face challenges in gaining market acceptance, especially if there are concerns about the quality or credibility of the underlying projects.

Reference link: https://gemini.google.com/

OFFSET CCCS (O-CCC)

Offset CCCs (O-CCC) are a type of carbon credit that specifically targets offsetting emissions. This means they are intended to neutralize or balance greenhouse gas emissions from one source by supporting projects that reduce emissions elsewhere.

Key characteristics of O-CCC:

Offsetting purpose:

O-CCC are explicitly designed to offset emissions, as opposed to other types of carbon credits that may have broader sustainability goals.

Verifiable emissions reductions:

The underlying projects must provide verifiable emissions reductions to ensure the credibility of the credits.

Compliance:

O-CCC can be used to comply with emissions regulations or voluntary carbon neutrality commitments.

Market-based: They are traded on carbon markets, allowing entities to purchase and retire them to offset their own emissions.

EXAMPLES OF O-CCC PROJECTS

Renewable energy: Investing in solar, wind, or hydroelectric power plants.

Energy efficiency: Improving energy efficiency in buildings or industrial processes.

Forest conservation: Protecting and restoring forests to sequester carbon.

Agricultural practices: Promoting sustainable agricultural practices that reduce emissions.

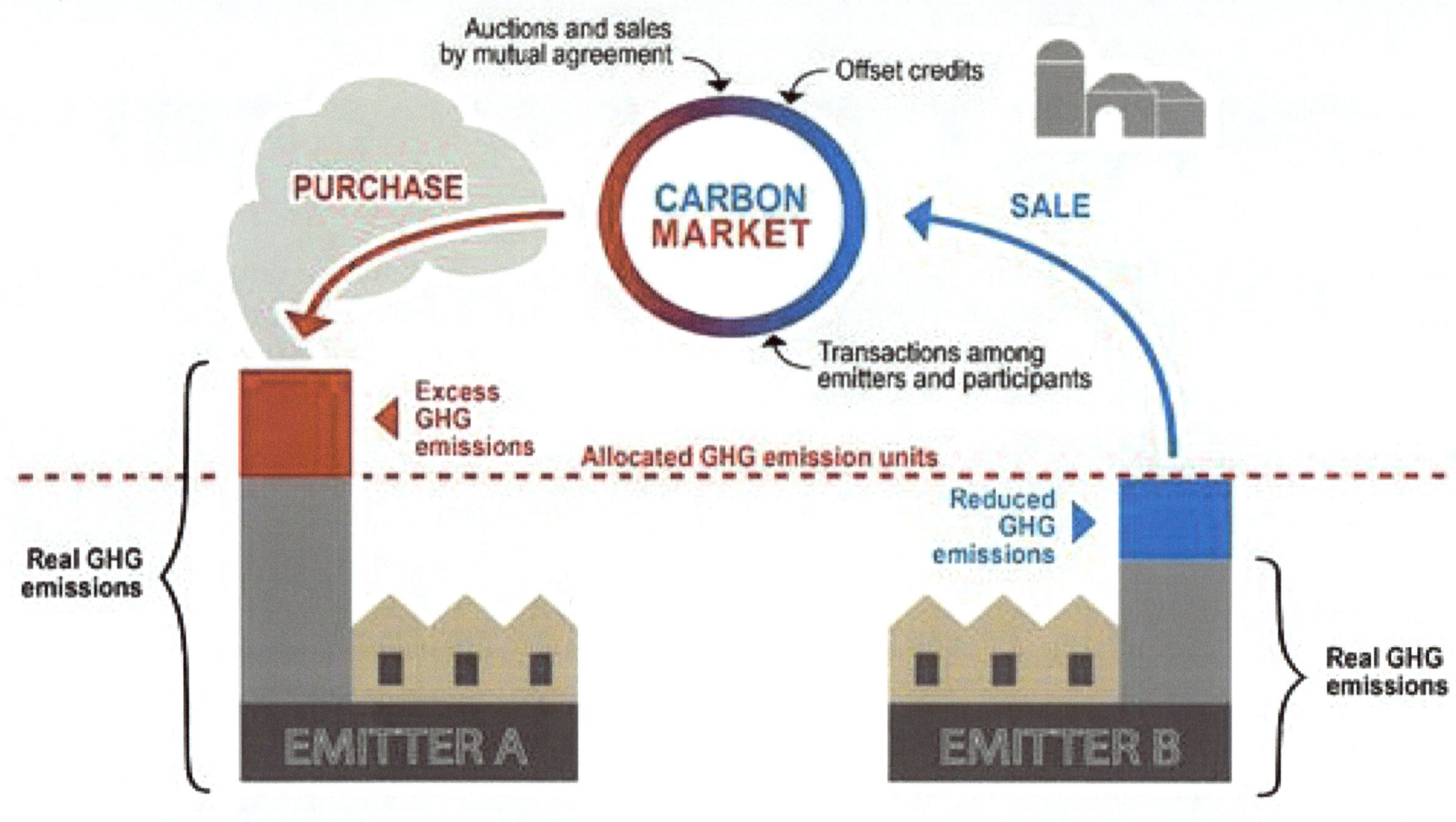

INDIAN CARBON MARKET (ICM)

The Indian Carbon Market (ICM) is an emerging framework aimed at reducing greenhouse gas (GHG) emissions in India by allowing companies to trade carbon credits. These credits represent a reduction of one metric ton of CO_2 equivalent, and companies can earn or purchase them based on their emissions performance. The market is being developed under the Carbon Credit Trading Scheme (CCTS), which was formalized through an amendment to the Energy Conservation Act in 2022.

The ICM is designed to help India meet its climate goals, including reducing the emissions intensity of its GDP by 45% by 2030, compared to 2005 levels, and achieving net-zero emissions by 2070. It will operate under a compliance mechanism where obligated entities must meet specific emissions targets. If they exceed these targets, they can trade their surplus credits; if they fall short, they must purchase additional credits to comply with regulations.

THE ICM WILL HAVE FOLLOWING BENEFITS:

It will help India lower the emissions intensity of its GDP by 45% by 2030 compared to the 2005 levels, thereby meeting its NDC target related to its global climate commitments.

ICM would help in decarbonising the commercial and industrial segments (in line with India's net zero by 2070).

it will give a fillip to energy transition due to its greater scope for covering the country's potential energy segments.

GHG emissions intensity targets and benchmarks would then be developed in sync with the domestic emissions trajectory, according to the climate goals.

Although the ICM would be regulated, it will offer flexibility to companies in hard-to-abate segments to augment their GHG emission efforts through carbon market credits. It will also create more awareness, change and innovation across hard-to-abate industries.

It could help attract finance and technology for sustainable projects that can generate carbon credits.

Source: http://drishtiias.com/images/uploads/1694860738_image1.jpg
Reference link: https://www.drishtiias.com/daily-updates/daily-news-editorials/carbon-markets-in-india-a-catalyst-for-green-growth

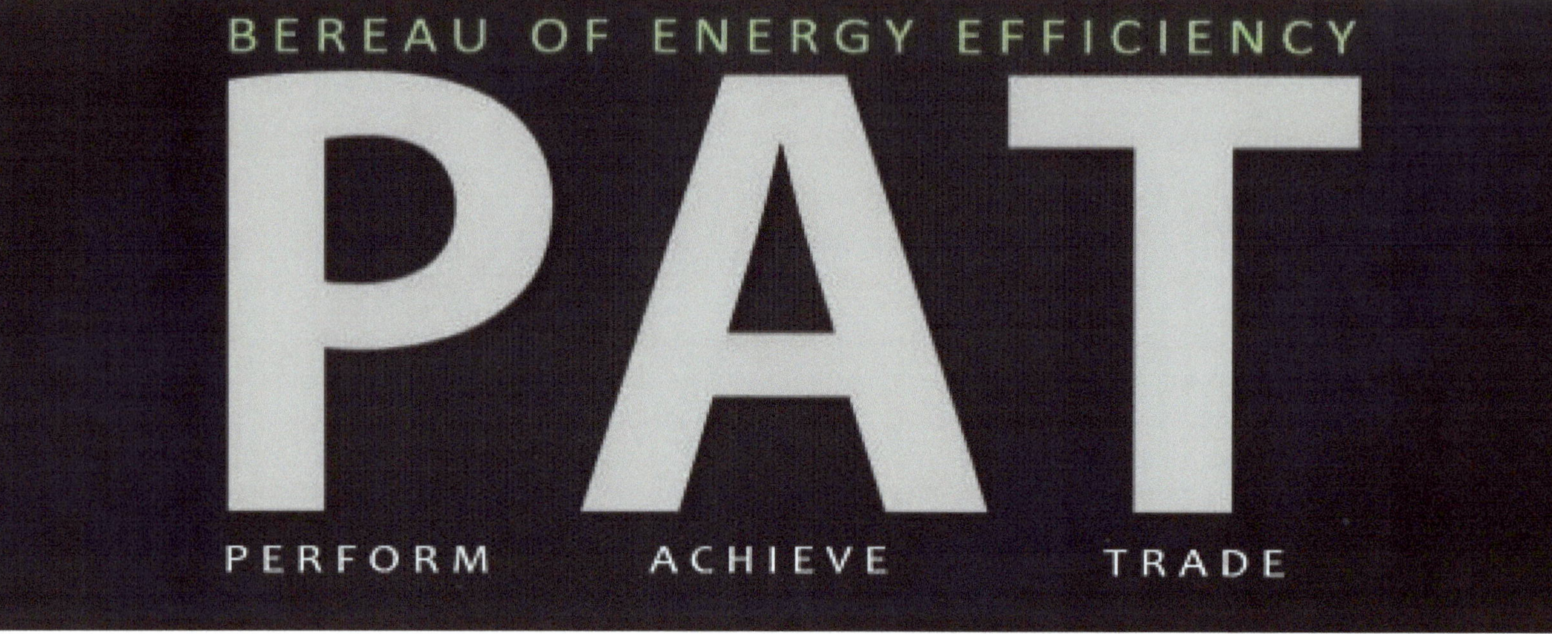

PERFORM ACHIEVE TRADE (PAT) SCHEME OF INDIA

Perform, Achieve and Trade (PAT) is a regulatory instrument to reduce Specific Energy Consumption in energy intensive industries, with an associated market-based mechanism to enhance the cost effectiveness through certification of excess energy saving which can be traded. PAT is a mechanism for improvements in energy efficiency of energy intensive industries. Specific high energy intensive industries are identified as Designated Consumers (DCs) within certain key sectors, who are required to appoint an energy manager, file energy consumption returns every year and conduct mandatory energy audits regularly.

The key tasks in the PAT mechanism is to set the methodology for deciding the Specific Energy Consumption (SEC) norms for each designated consumers in the baseline year and in the target years, devise verification process for SEC, finding ways of issuing the Energy Savings Certificates, operationalization of the trading process for ESCerts in addition to the compliance and reconciliation process for ESCerts.

Background

Recognizing the need to maintain a high growth rate for increasing the living standards of vast majority of people and reducing the vulnerability to adverse impacts of Climate Change, National Action Plan on Climate Change (NAPCC) was released in June 2008. NAPCC outlined eight national missions representing long-term and integrated strategies for achieving key goals in the context of climate change. The National Mission for Enhanced Energy Efficiency (NMEEE) is one of the eight missions released under the NAPCC. The implementation plans for NMEEE were entrusted with the Ministry of Power and Bureau of Energy Efficiency. NMEEE unrolled the following four initiatives:

Perform Achieve and Trade Scheme (PAT)

Market Transformation for Energy Efficiency (MTEE)

Energy Efficiency Financing Platform (EEFP)

Framework for Energy Efficient Economic Development (FEEED)

Creation of the PAT mechanism comes from the provisions of the Energy Conservation Act, 2001 which also empowers the Central Government to notify energy intensive industries as listed out in the Schedule to the Act, as Designated Consumers (DCs).

Bureau of Energy Efficiency had conducted sector specific studies for setting of energy consumption norms and standards. The studies showed that there is a wide bandwidth of SEC within an industrial sector that indicated large energy-savings potential in the sector. For setting targets for Designated Consumers, BEE had carried out background work to enable designing of a transparent, flexible, efficient and robust system for the PAT mechanism. BEE also consulted key stakeholders like Designated Consumers, Energy Auditors/ Managers, Industry Associations, Academics, etc. and solicited comments while framing the complete mechanism of PAT Scheme.

Source: https://gshindi.com/sites/default/files/2018-05/PAT.jpg & https://powerline.net.in/wp-content/uploads/2018/03/42.jpg
Reference link: https://beeindia.gov.in/en/programmes/perform-achieve-and-trade-pat#:~:text=Perform%2C%20Achieve%20and%20Trade%20(PAT)%20is%20a%20regulatory%20instrument,saving%20which%20can%20be%20traded.

ACCREDITED CARBON
VERIFICATION AGENCY (ACV)

An Accredited Carbon Verification Agency (ACV) is a third-party organization that validates and verifies greenhouse gas (GHG) emissions reductions or removals projects. These agencies play a crucial role in ensuring the integrity and credibility of carbon offset markets.

KEY FUNCTIONS OF AN ACV:

Verification: Assessing the accuracy and completeness of data related to GHG emissions reductions or removals.

Validation: Confirming that the project methodology used to calculate emissions reductions or removals is sound and aligns with established standards.

Certification: Issuing certificates that confirm the verified and validated emissions reductions or removals.

ACCREDITATION

ACVs are typically accredited by recognized bodies, which ensure that they meet specific standards and requirements. This accreditation process provides assurance of their competence, impartiality, and adherence to ethical principles.

EXAMPLES OF ACVS:

While the specific ACVs operating in your region may vary, here are some well-known international organizations that often provide accreditation or verification services:

Verra: A global non-profit organization that develops and maintains standards for voluntary carbon markets.

Gold Standard: A non-profit organization that sets environmental and social standards for carbon offset projects.

American Carbon Registry (ACR): A non-profit organization that provides voluntary carbon offset registry services in the United States.

Climate Action Reserve: A non-profit organization that provides carbon offset verification and registry services in California.

CO₂ TRACKING

WHAT IS CARBON TRACKING?

Carbon tracking is a method for organizations to calculate and track carbon emissions released from both direct and indirect sources. By tracking carbon emissions, companies can measure the carbon impact of their business activities and pinpoint carbon hotspots', which they should prioritize to make emissions reductions. Carbon tracking should be a continuous process, rather than just a one-off measurement, to measure, set targets and track progress against them. This is critical for the accountability and reporting that stakeholders now demand.

WHY DO BUSINESSES UNDERTAKE CARBON TRACKING?

● Businesses are measuring their carbon footprints to meet regulatory requirements, better understand their impact and focus efforts on where they can make the most meaningful changes.

● Manufacturers and commodity traders operating in high-emitting sectors such as metals, energy and transport are recognizing the business benefits of identifying carbon hotspots across the supply chain. Financial institutions are also increasingly engaging in carbon tracking to have an overview of the emissions of the companies they are investing in, and the trades they are financing.

● Differing from carbon reporting — which comes after the tracking process, and is a way to present data collected on GHG emissions — carbon tracking is an important first step for companies to measure and understand their environmental impact more fully.

HOW DOES CARBON TRACKING WORK?

Worldwide, there are a range of global carbon trackers that measure GHG emissions in different ways, including:

MEASURING GOVERNMENTAL CLIMATE ACTION

The Climate Action Tracker (CAT) is an independent scientific platform which tracks governmental climate action and compares it to the Paris Agreement's aim to keep warming "well below 2°C, and pursue efforts to limit warming to 1.5°C". The CAT covers around 85% of global carbon emissions and around 70% of the global population.

MEASURING GLOBAL EMISSIONS

Carbon Tracker is a CO_2 measurement and modelling system developed by the US National Oceanic and Atmospheric Administration (NOAA) agency, which tracks emissions sources (and sinks, which remove carbon from the atmosphere) from 81 sites around the world, with the help of global collaborators who provide atmospheric CO_2 observations.

These trackers help quantify global emissions and country-level commitments, and are therefore important for policymakers, scientists and a range of other stakeholders.

However, they do not provide information on who is responsible for these emissions, nor do they link them to specific sources. Limitations and data gaps on emitters and their sources remain, while national inventories have also been found to underestimate many countries' GHG emissions.

Asset-level emissions tracking, where emissions are associated with the facilities responsible for those emissions, is a critical piece of information that global trackers are still working to solve. For companies to reach net zero by 2050, they will need detailed data on their own GHG emissions, as well as those of their suppliers and customers. Accurately measuring their carbon emissions will enable businesses to lower their carbon footprint, across the value chain.

CARBON CREDIT TOKENIZATION

Carbon credit tokenization is the process of creating digital tokens on Blockchain that represent carbon credits. One can easily buy, sell, and trade these credits like other digital assets in the Blockchain space.

STEPS FOR TOKENIZING CARBON CREDITS

The process of tokenizing carbon credit involves a number of stages

- The process begins with simply verifying and quantifying the carbon emission reduction achieved by a particular project.
- Once the data is verified, it is then recorded on Blockchain. The data stored on the Blockchain becomes immutable that restricts the changes by the unauthorized users. It will also remain transparent to everyone in the network.
- Tokens representing carbon credits are further issued via smart contracts on Blockchain. Smart contracts automate the issuance and transfer of carbon credit tokens that not only save time and resources but also make the transactions more efficient and reliable.
- Once carbon credits are tokenized, these digital assets can be easily bought, sold, and traded.

BENEFITS OF CARBON CREDIT TOKENIZATION

Carbon credit token development is a revolutionary step to streamline the voluntary carbon market. Below, we have covered the benefits of carbon credit tokenization, highlighting its importance in the carbon market.

FRACTIONALIZATION

Carbon credit tokenization also promotes the fractionalization of credits. Tokenized carbon credits can be divided into units smaller than one metric tonne just like a currency. It helps small-scale carbon projects to offset their carbon footprint.

BUILD A HEALTHIER MARKET

Carbon credit tokenization reduces the counterparty risks and helps build a more efficient market for individuals and organizations. Trades can be settled without any delay. Buying and selling carbon credit also becomes convenient with carbon credit token development as there is no need to set up an account or get approval. Retirement can also be done in a few minutes which usually takes a few months.

LIQUID MARKETPLACES

Liquidity refers to the ease with which an asset can be traded for cash. Carbon credit token development also helps build liquidity for the illiquid carbon market. With Blockchain technology, aggregating the carbon credits with similar attributes is easier that further helps build more liquidity and set a fair price for the asset.

EFFICIENCY & DISINTERMEDIATION

Intermediaries like middlemen or brokers in the existing voluntary carbon market often educate buyers, create carbon credits, and retire them on behalf of the purchaser. However, they charge a high amount of fees for their services. Additionally, they buy carbon credits from developing nations at a cheap price and often sell them at an inflated price (sometimes it is three times the original value).

Carbon credit token development eliminates the need for intermediaries which are the major key drivers of costly and delayed transactions. Blockchain reduces the risk of fraud to a great extent as there is no need to trust the third party for the transactions.

ELIMINATE DOUBLE COUNTING

Double counting is one of the common issues in the carbon credit tokenization process. Double counting claims refer to different sustainability claims levied on the same carbon credit. The decentralized and distributed public ledger of Blockchain also makes it easier to locate double-counting claims that are usually one of the major issues in the voluntary carbon market.

NATIONAL ADA

UNEP assists countries all over the world in their efforts to create National Adaptation Plans (NAPs). The NAP process seeks to identify medium- and long-term adaptation needs, informed by the latest climate science. Once major vulnerabilities to climate change have been identified, the NAP process develops strategies to address them.

The NAP approach was established under the Cancun Adaptation Framework (CAF) and re-emphasized in the Paris Agreement. Crucially, NAPs follow a continuous iterative process that is country-driven, participatory and transparent.

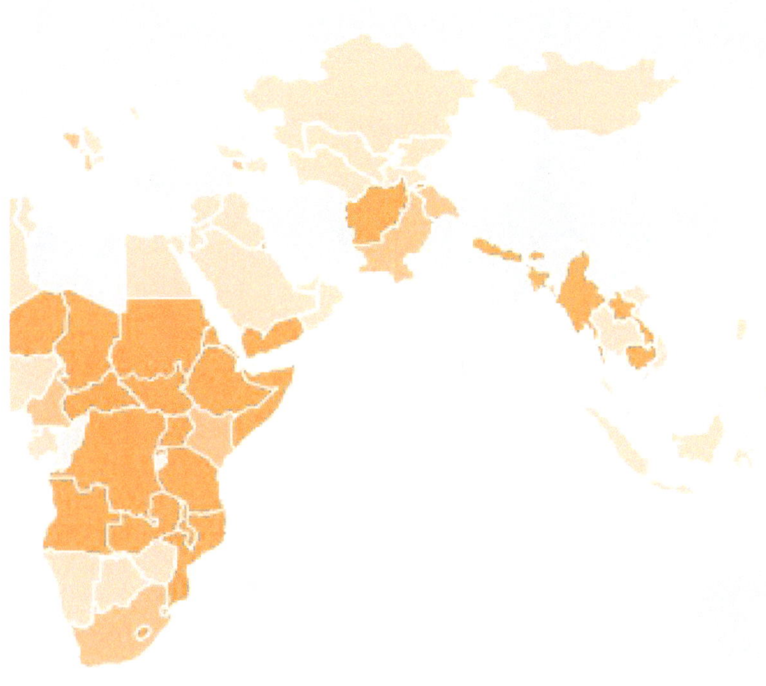

The two overarching objectives of NAPs are to:

- Reduce vulnerability to the impacts of climate change by building adaptive capacity and resilience;
- Facilitate the integration of climate change adaptation, in a coherent manner, into relevant new and existing policies, programmes and activities, in particular development planning processes and strategies, within all relevant sectors and at different levels, as appropriate.

Countries can utilize the NAP process and its outcomes to update and improve the adaptation elements of the Nationally Determined Contributions (NDCs), a central part of the Paris Agreement.

untry Parties
nitted a NAP 

Least developed country Parties
that have submitted a NAP 

There are three avenues through which UNEP supports the development of NAPs:

THE NATIONAL ADAPTATION PLAN GLOBAL SUPPORT PROGRAMME (NAP-GSP) – This project, funded by the Global Environment Facility, was run jointly with UNDP between 2013 and 2021 and provided support to over 45 countries.

ONE-ON-ONE COUNTRY SUPPORT PROGRAMMES –

UNEP has more than 23 ongoing projects where one-on-one support is provided to countries to advance their NAPs. These NAP projects are funded by either the Green Climate Fund (GCF) or the Global Environment Facility (GEF).

KNOWLEDGE SHARING -

UNEP generates a wide range of science, analyses and knowledge sharing events to support countries in their efforts to build climate resilience. For example, UNEP published its Guidelines for Integrating Ecosystem-based Adaptation into National Adaptation Plans, which has now been transformed into an online course. For more climate adaptation resources, visit us here. Map of developing and least developed country Parties that had submitted national adaptation plans or initiated the process to formulate and implement NAPs (lighter color) as at 11 November 2023.

Blue economy is an economic term linked to exploitation and conservation of the maritime environment and is sometimes used as a synonym for "sustainable ocean-based economy". There is, however, no consensus on the exact definition and the field of application depends on organization that uses it. The UN first introduced "blue economy" at a conference in 2012 and underlined sustainable management, based on the argument that marine ecosystems are more productive when they are healthy. This is backed up by scientific findings, showing that the earth's resources are limited and that greenhouse gases are damaging the planet. Furthermore, pollution, unsustainable fishing, habitat destruction etc. harm the marine life and are increasing day by day. The UN specifies Blue Economy as a range of economic activities related to oceans, seas and coastal areas, and whether these activities are sustainable and socially equitable. An important key point of Blue Economy is sustainable fishing, ocean health, wildlife, and stopping pollution. The UN iterates that the Blue Economy should "promote economic growth, social inclusion, and the preservation or improvement of livelihoods while at the same time ensuring environmental sustainability of the oceans and coastal areas". This points out the importance of global cooperation across borders and sectors. This also indicates that governments, organizations and decisionmakers need to join forces to ensure that their policies won't undermine each other.

Source: https://unfccc.int/topics/adaptation-and-resilience/workstreams/national-adaptation-plans
Reference link: https://www.unep.org/topics/climate-action/adaptation/national-adaptation-plans#:~:text=The%20NAP%20process%20seeks%20to,develops%20strategies%20to%20address%20them.

The use of the seas, the oceans and the coastal areas has accelerated the past years. The OECD describes the ocean as the next great economic frontier as it holds potential for wealth and economic growth, employment and innovation. And while the economy includes existing businesses such as fisheries, coastal tourism and shipping, it also focuses on the development of new emerging sectors that were next to non-existent 20 years ago e.g. blue carbon sequestration, marine energy and biotechnology; sectorial activities that create potential and opportunities for training and employment, but also fight climate change. Benefits of blue economy: create green energy and fight climate change.

Blue Economy has the power to obtain better governance of marine ecosystems, lower emissions, a more just health standard and be a player in fighting climate change. In the recent years, emerging sectors within energy have grown exponentially, and oceans are popular sites for renewable energy. Alternative energy sources such as wind energy, hydropower and tidal energy are fitting for marine environments. Especially offshore wind (including floating wind turbines) is fast growing and has around for many years – the first offshore wind park erected in 1991 in Denmark, and the quantity of offshore wind farms was 162 in 2020, according to WFO. The report Offshore Wind Outlook 2019 by the International Energy Agency (IEA), offshore wind power has the potential to generate more than 18 times the global electricity demand today. Wind farms require specific professions and therefore create jobs in construction, maintenance and administration. Offshore wind energy is only one example of benefits of Blue Economy. Others are offshore aquaculture (an emerging approach to fish farming), wave and tidal energy, seabed mining and blue biotechnology, which uses, among others, shellfish, bacteria and algae for development in health care and energy production. Moreover, existing industries, such as shipping and tourism, have potential of growing and become greener with new technologies.

To support the Blue Economy, both the European Union and the United Nations have developed a long-term strategy that aims to support facilitate sustainable ocean-based economic benefits by implementing climate-resilient and inclusive blue economy policies that reduce human impact. Some countries have also taken it upon themselves to implement strategies and policies that support the idea of Blue Economy. Among these are Denmark and Norway that have a clear focus on the shipping industry.

Source: https://acecor.ucc.edu.gh/sites/default/files/styles/normal_size/public/2019-07/Blue%20Economy.jpg?itok=Ew-5eupE & https://www.teriin.org/sites/default/files/2021-03/ocean-based-livlihood-og.jpg Reference link: https://unric.org/en/blue-economy-oceans-as-the-next-great-economic-frontier/

SUSTAINABLE INVESTING

Sustainable investing refers to a range of practices in which investors aim to achieve financial returns while promoting long-term environmental or social value. Combining traditional investment approaches with environmental, social, and corporate governance (ESG) insights has led to investors generating more comprehensive analyses and making better investment decisions. Sustainable investing ensures firms aren't judged solely on short-term financial gains but on a broader picture of what and how they contribute to society. Investors must think critically about investments' potential impacts as they relate to environmental, political, and societal landscapes.

SUSTAINABLE INVESTING STRATEGIES

While most sustainable initiatives share the same end goals, not all investors share the same motivations. Therefore, there are several strategies business leaders can leverage when investing sustainably, including:

● Negative/exclusionary screening: Excludes specific sectors, companies, or practices from a fund or portfolio based on ESG criteria

● Positive/best-in-class screening: Encompasses investments in sectors, companies, or projects selected from a defined universe for positive ESG performance compared to industry peers

An investment's sustainability impact is evaluated using ESG factors. Here's a breakdown of what an ESG score typically consists of:

Environmental:

A company's impact on the environment, such as its carbon footprint, waste, water use and conservation, and the clean technology it uses and creates in its supply chain.

Social:

A company or fund's impact on society and how it advocates for social good and change. Analysts closely examine its involvement and stances on social issues, such as human rights, racial diversity within hiring and inclusion programs, employees' health and safety, and community engagement.

Governance:

How an exchange-traded fund (ETF) or company is managed or "governed" for driving positive change. It encompasses reviewing the quality of its management and board, executive compensation and diversity, shareholder rights, overall transparency and disclosure, anti-corruption, and corporate political contributions.

ESG factors are a significant consideration in sustainable investing, but there are additional strategies investors can use:

ACTIVIST INVESTING:

Buying equity in a company for the purpose of changing how it operates. Investment decisions are based on moral values or causes that companies and their leaders care deeply about. For example, individuals who strongly care about global warming may invest in a company that drives environmental change.

IMPACT INVESTING:

Targeted investments aimed at solving social or environmental problems. It includes community investing, where capital is directed to traditionally underserved individuals or communities and financing is provided to businesses with clear social or environmental purposes. While impact investing is traditionally referred to as a private market strategy, there are now public market funds that identify as impact investors.

SOCIALLY RESPONSIBLE INVESTING

Socially responsible investing (SRI), also known as social investment, is an investment that is considered socially responsible due to the nature of the business the company conducts. A common theme for socially responsible investments is socially conscious investing. Socially responsible investments can be made into individual companies with good social value, or through a socially conscious mutual fund or exchange-traded fund (ETF).

Understanding Socially Responsible Investment (SRI)

Socially responsible investments—known as conscious capitalism—include eschewing investments in companies that produce or sell addictive substances or activities (like alcohol, gambling, and tobacco) in favour of seeking out companies that are engaged in social justice, environmental sustainability, and alternative energy/clean technology efforts.

In recent history, socially conscious investing has been growing into a widely-followed practice, as there are dozens of new funds and pooled investment vehicles available for retail investors. Mutual funds and ETFs provide an added advantage in that investors can gain exposure to multiple companies across many sectors with a single investment. However, investors should read carefully through fund prospectuses to determine the exact philosophies being employed by fund managers, along with the potential profitability of these investments.

There are two inherent goals of socially responsible investing: social impact and financial gain. The two do not necessarily have to go hand in hand; just because an investment touts itself as socially responsible doesn't mean that it will provide investors with a good return and the promise of a good return is far from an assurance that the nature of the company involved is socially conscious. An investor must still assess the financial outlook of the investment while trying to gauge its social value.

Example of Socially Responsible Investing:
One example of socially responsible investing is community investing, which goes directly toward organizations that both have a track record of social responsibility through helping the community, and have been unable to garner funds from other sources such as banks and financial institutions. The funds allow these organizations to provide services to their communities, such as affordable housing and loans. The goal is to improve the quality of the community by reducing its dependency on government assistance such as welfare, which in turn has a positive impact on the community's economy.

Source: https://sri360.com/wp-content/uploads/2022/10/Main-image-resized-1024x683.jpg https://serudsindia.org/wp-content/uploads/2024/03/socially-responsible-investing.jpg
Reference link: https://www.investopedia.com/terms/s/sri.asp

SUSTAINABLE FINANCE

Sustainable Finance is the process of taking due account of environmental, social and governance (ESG) considerations when making investment decisions in the financial sector, leading to increased longer-term investments into sustainable economic activities and projects (European Commission). It has become a powerful movement led by regulators, institutional investors and asset managers globally. Sustainability, however, is a complex and evolving topic. The World Bank Group long-term finance unit has been at the forefront of promoting Sustainable Finance globally – though data provision, analytical work, instrument design and technical assistance to support regulators and investors in our client countries to 'green' their financial systems.

The team's work on sustainable finance contributes to several initiatives - namely:

1. The Global Program on Sustainability which promotes the use of high quality-data and analysis on natural capital, ecosystem services and sustainability to better inform decisions made by governments, the private sector and financial institutions. The GPS program consists of 3 key pillars:

2. The Sovereign ESG Data Portal is part of the work supported by the Global Program on Sustainability (GPS), which aims to provide governments and investors with information and tools that improve their understanding of sustainability criteria, including through natural capital accounting.

Pillar 1: Information-improving global measurements of natural capital and ecosystem services.

Pillar 3: Incentives- Promoting research on how environmental Factors impact risk and financial return in fixed income markets.

Pillar 2: Building countries capacity to produce and use natural capital accounting for policy and planning decisions. It currently works with 18 countries to measure and value natural resources.

3. Climate Support Facility is a new flagship trust fund which was launched on December 10, 2020. The facility manages funding provided under a Green Recovery Initiative aimed at helping countries building a low-carbon, climate-resilient recovery from COVID 19. Germany, the United Kingdom and Austria are its first contributors. The facility supports technical assistance and advisory services.

4. IFC Edge. An innovation of IFC, a member of the World Bank Group, EDGE ("Excellence in Design for Greater Efficiencies") provides market leaders with the opportunity to gain a competitive advantage by differentiating their products and adding value to the lives of their customers. EDGE brings speed, market intelligence and an investment focus to the next generation of green building certification in more than 170 countries. IFC created EDGE to respond to the need for a measurable and credible solution to prove the business case for building green and unlock financial investment.

5. J-CAP is a five-year program initially focused on six priority countries and one sub-region: Bangladesh, Indonesia, Kenya, Morocco, Peru, Vietnam, and the West African Economic & Monetary Union. Going forward, the program will also work with other countries, including Argentina, to support local capital markets development. Under the program, country-specific action plans have been produced that mobilize the World Bank's technical assistance alongside IFC's demonstration transactions and local currency solutions. Support for green bond issuance and market development, as well as green regulatory frameworks are a core part of the program.

COMMON BUT DIFFERENTIATED
RESPONSIBILITIES AND RESPECTIVE CAPABILITIES

Common but Differentiated Responsibilities and Respective Capabilities (CBDR–RC) is a principle within the United Nations Framework Convention on Climate Change (UNFCCC) that acknowledges the different capabilities and differing responsibilities of individual countries in addressing climate change. The principle of CBDR–RC is enshrined in the 1992 UNFCCC treaty, which was ratified by all participating countries. The text of the convention reads: "… the global nature of climate change calls for the widest possible cooperation by all countries and their participation in an effective and appropriate international response, in accordance with their common but differentiated responsibilities and respective capabilities and their social and economic conditions."

CBDR-RC has served as a guiding principle as well as a source of contention in the UN climate negotiations. Reflecting CBDR-RC, the Convention divided countries into "Annex I" and "non-Annex I," the former generally referring to developed countries and the latter to developing countries. Under the Convention Annex I countries have a greater mitigation role than non-Annex-I countries. Since 1992 countries like China have gained new capabilities while maintaining relatively low per capita emissions, and tensions about the defined lines of the Annex I and non-Annex I countries have arisen. CBDR-RC and the annex classifications were codified in the 1997 Kyoto Protocol, and Annex I country emissions reductions were legally bound.

A primary driver for the failure of the U.S. to ratify the Kyoto Protocol was the domestic concern that middle-income developing countries were not required to take action to address their greenhouse gas (GHG) emissions despite their growing capability. In the years following the 1992 treaty, the trajectory of emissions in populous developing countries also drew attention. Fossil fuel–based development by heavily populated developing

countries would prevent stabilization of GHG concentrations – the agreed upon "ultimate objective" of the UNFCCC – because much of the global emissions budget has already been exhausted by emissions from developed countries. Controversy ensued over the question of responsibility for the costs entailed in switching to a sustainable development path, particularly for large but poor countries with very low per-capita emissions and very little access to finance. However, in more recent UNFCCC agreements – starting with Durban in 2011 – Parties have changed their position to allow for countries to individually determine their "contribution" to addressing GHG emissions. This new climate agreement is to be "applicable to all," and approaches differentiation through the implementation of a bottom–up scheme to determine a global effort.

CBDR-RC remains a sticking point, as does the role of equity (historic versus current responsibility for climate change), the role of Annexes, and the role each country should play in in UNFCCC climate negotiations. In the 2014 negotiations in Lima, Parties agreed on a new phrase, 'common but differentiated responsibilities and respective capabilities, in light of different national circumstances,' perhaps hinting at how an agreement in Paris would address the issue.

Source: https://miro.medium.com/v2/resize:fit:359/1*GtoTeB-CT8wgoC-0q3lBzg.png Reference link: https://climatenexus.org/climate-change-news/common-but-differentiated-responsibilities-and-respective-capabilities-cbdr-rc/

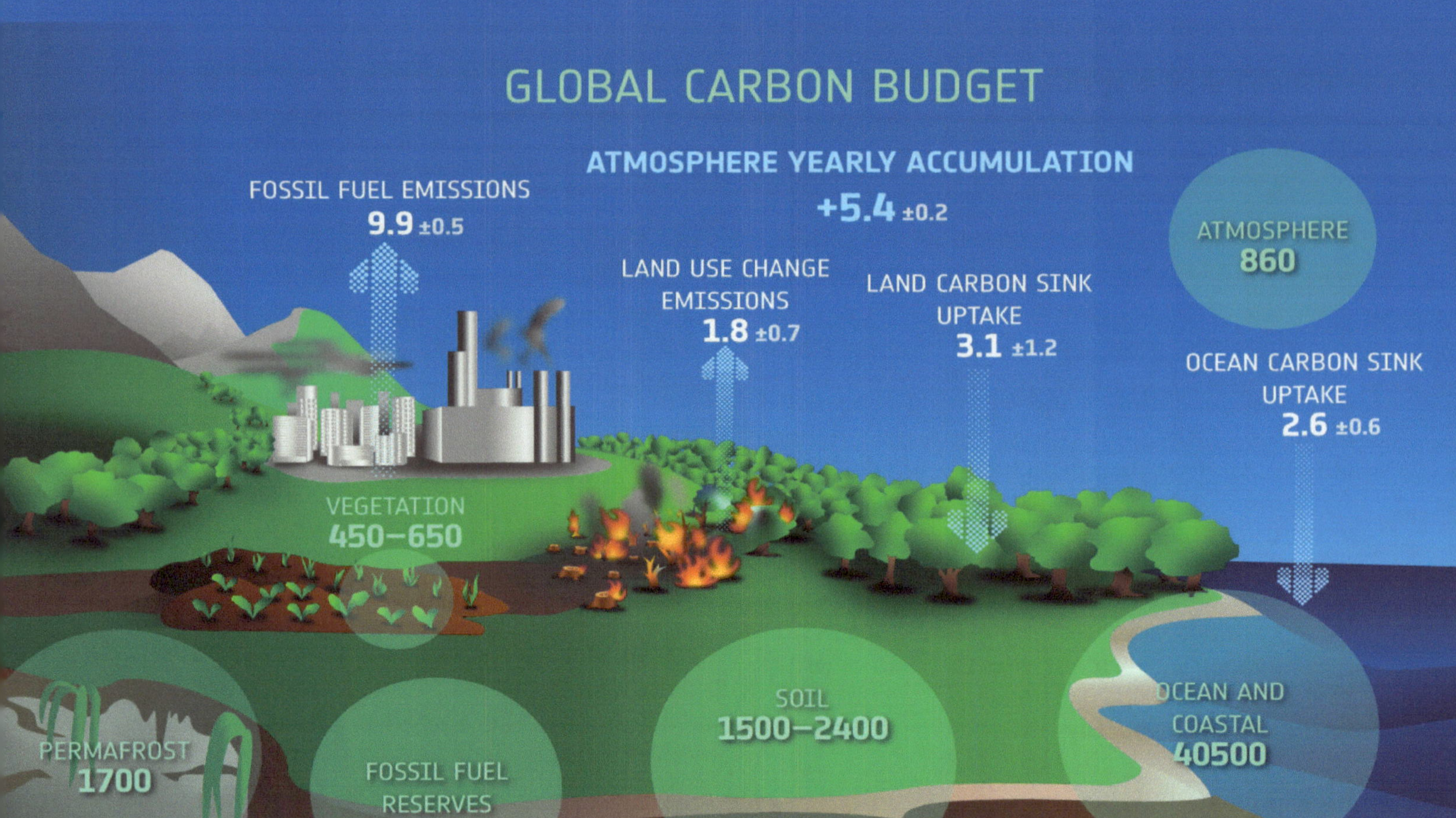

GLOBAL CARBON BUDGET

A carbon budget is a concept used in climate policy to help set emissions reduction targets in a fair and effective way. It examines the "maximum amount of cumulative net global anthropogenic carbon dioxide (CO_2) emissions that would result in limiting global warming to a given level". It can be expressed relative to the pre-industrial period (the year 1750). In this case, it is the total carbon budget. Or it can be expressed from a recent specified date onwards. In that case it is the remaining carbon budget.

A carbon budget that will keep global warming below a specified temperature limit is also called an emissions budget or quota, or allowable emissions. Apart from limiting the global temperature increase, another objective of such an emissions budget can be to limit sea level rise.Scientists combine estimates of various contributing factors to calculate the carbon budget. The estimates take into account the available scientific evidence as well as value judgments or choices.

Global carbon budgets can be further sub-divided into national emissions budgets. This can help countries set their own emission goals. Emissions budgets indicate a finite amount of carbon dioxide that can be emitted over time, before resulting in dangerous levels of global warming. The change in global temperature is independent of the source of these emissions, and is largely independent of the timing of these emissions.

To translate global carbon budgets to the country level, a set of value judgments have to be made on how to distribute the remaining carbon budget over all the different countries. This should take into account aspects of equity and fairness between countries as well as other methodological choices. There are many differences between nations, such as population size, level of industrialisation, historic emissions, and mitigation capabilities. For this reason, scientists are attempting to allocate global carbon budgets among countries using various principles of equity.

Source: https://www.esa.int/var/esa/storage/images/esa_multimedia/images/2021/11/global_carbon_budget/23761660-1-eng-GB/Global_carbon_budget_pillars.jpg Reference link: https://en.wikipedia.org/wiki/Carbon_budget

GLOBAL STOCKTAKE (GST)

What is the Global Stocktake?

The global stocktake of the Paris Agreement (GST) is a process for taking stock of the implementation of the Paris Agreement with the aim to assess the world's collective progress towards achieving the purpose of the agreement and its long-term goals (Article 14). The first stocktake got underway at the UN Climate Change Conference in Glasgow last November (COP26) and will conclude at COP28 in 2023. Each stocktake is a two-year process that happens every five years. The first global stocktake is critical to assessing collective progress under the Paris Agreement and addressing opportunities for enhanced action and support. The highly participatory structure of the GST opens the door for stakeholders to provide vital input and contribute to our understanding of global efforts and priority actions.

On March 30 2022, participants of the MENA Climate Week (MENACW2022) heard updates on the first Global stocktake and discussed progress, challenges, and opportunities in responding to the Paris Agreement in the MENA region.

"The global stocktake will provide critical information for countries and stakeholders to see what progress has been made on meeting the Paris Agreement goals, as well as identify any remaining gaps and opportunities for increased action," said Joanna Post, Programme Officer with UN Climate Change's Intergovernmental Support and Collective Progress Division. "To put it simply – the global stocktake will help spur countries to step up climate action to avoid the worst impacts of climate change."

Why the Global Stocktake matters

The Paris Agreement calls on each country to set its own plan to cut emissions and adapt to climate impacts. It also established a process for countries to continually strengthen their national climate plans. These national climate change plans are formally known as Nationally Determined Contributions (NDCs).

To hold themselves accountable, countries agreed to regularly report on and review their individual efforts and to take stock of their collective progress. The Paris Agreement created the global stocktaking process, which follows a five-year cycle.

The global stocktake will help national governments see what they have achieved so far in implementing their climate plans, identify what still needs to be done to meet their targets, and highlight opportunities to increase their ambition on climate action.

The global stocktake is critically important because the international community has yet to live up to its commitments and climate action has yet to reflect deep transformations needed across all sectors to build a resilient future.

GLOBAL DECARBONISATION ACCELERATOR (GDA)

At the World Climate Action Summit, COP28 President Dr. Sultan Al Jaber unveiled the Global Decarbonization Accelerator (GDA), a series of landmark initiatives designed to speed up the energy transition and drastically reduce global emissions. The GDA is focused on three key pillars: rapidly scaling the energy system of tomorrow; decarbonizing the energy system of today; and targeting methane and other non-CO_2 greenhouse gases (GHGs). It is a comprehensive plan for system wide change, addressing the demand and the supply of energy at the same time.

The GDA has been informed by the thinking of key stakeholders, including the international organisations, governments and policy makers, NGOs, and CEOs from every industrial sector. Commentating on the launch of the GDA, Dr. Al Jaber said: "The world does not work without energy. Yet the world will break down if we do not fix energies we use today, mitigate their emissions at a gigaton scale, and rapidly transition to zero carbon alternatives. That is why the COP28 Presidency has launched the Global Decarbonization Accelerator."

Source: https://climateinsider.com/wp-content/uploads/2023/12/Screen-Shot-2023-12-05-at-12.41.57-AM-1024x573.png
Reference link: https://www.cop28.com/en/news/2023/12/COP28-Presidency-launches-landmark-initiatives-accelerating-the-energy-transition

Rapidly scaling the energy system of tomorrow

Today 116 countries have signed the Global Renewables and Energy Efficiency Pledge as of today, agreeing to triple worldwide installed renewable energy generation capacity to at least 11,000 gigawatts and to double the global average annual rate of energy efficiency improvements from around 2 percent to more than 4 percent ever year until 2030. Through the UAE Hydrogen Declaration of Intent, 27 countries have agreed to endorse a global certification standard and to recognize existing certification schemes, helping to unlock global trade in low-carbon hydrogen.

Decarbonizing the energy system of today

Under the GDA, 50 companies, representing over 40 percent of global oil production have signed on to the Oil and Gas Decarbonization Charter (OGDC), committing to zero methane emissions and ending routine flaring by 2030, and to total net-zero operations by 2050 at the latest. Over 29 National Oil Companies (NOCs) have committed to the Charter – the largest ever number of NOCs to sign up to a decarbonization pledge. The OGDC is an important step towards the industry increasing actions aligned with the aims of the Paris Agreement.

Methane and other non-CO_2 greenhouse gases

The third pillar of the GDA will addresses methane and other non-CO_2 greenhouse gases through economy-wide methane-emission reduction. In support of this more than \$1 billion will be mobilized for methane abatement projects, with additional information to be released on 5 December at the COP28 Energy Thematic Day. The GDA also covers the Global Cooling Pledge, which targets substantially reducing global cooling emissions by 68 percent by 2050. Such emissions account for seven percent of the global total, a figure expected to triple as more nations adopt air-conditioning. As of today, 52 countries have signed the Pledge.

Signatories to the Charter agree to target a number of key actions, including:

Investing in the energy system of the future including renewables, low-carbon fuels and negative emissions technologies.

Increasing transparency, including enhancing measurement, monitoring, reporting and independent verification of GHG emissions and their performance and progress in reducing emissions.

Increasing alignment with broader industry best practices to accelerate decarbonization of operations and aspire to implement current best practices by 2030 to collectively reduce emission intensity.

Reducing energy poverty and providing secure and affordable energy to support the development of all economies.

GLOBAL RENEWABLES AND ENERGY EFFICIENCY PLEDGE (COP28)

We, Heads of State and Governments as the Participants in the COP28 Global Renewables and Energy Efficiency Pledge:

● Recognizing that, to ensure that the global community meets the collective goal of the Paris Agreement to keep warming well below 2°C while pursuing efforts to limit warming to 1.5°C, the pace and scale of deployment of renewables and energy efficiency must increase significantly between now and 2030, propelling the global move towards energy systems free of unabated fossil fuels well ahead of and by mid-century at the latest.

● Noting that the International Energy Agency and the International Renewable Energy Agency forecast that, to limit warming to 1.5°C, the world requires three times more renewable energy capacity by 2030, or at least 11,000 GW, and must double the global average annual rate of energy efficiency improvements from around 2% to over 4% every year until 2030.

● Recognizing that energy is inextricably linked to all the UN Sustainable Development Goals, and transforming the world's energy systems will create new jobs, enhance lives and livelihoods, and empower people, communities, and societies.

● Recognizing the critical contribution of renewables and energy efficiency to the achievement of the UN Sustainable Development Goal 7 for "affordable, reliable, sustainable and modern energy for all", with nearly 800 million people globally without access to electricity, nearly 600 million of whom are in Africa.

● Recognizing that, according to the IEA and IPCC, in order to meet the Paris Agreement goal, renewables deployment must be accompanied in this decade by a rapid increase of energy efficiency improvements and the phase down of unabated coal power, in particular ending the continued investment in unabated new coal-fired power plants, which is incompatible with efforts to limit warming to 1.5°C.

● Noting the September 2023 G20 Leaders Declaration to "pursue and encourage efforts to triple renewable energy capacity globally through existing targets and policies, as well as demonstrate similar ambition with respect to other zero and low-emission technologies, including abatement and removal technologies, in line with national circumstances by 2030".

● Recognizing that, particularly in the post-2030 period, an increasingly diversified portfolio of technologies will be market-ready and available at scale to decarbonize the energy sector, with a critical role for renewables, energy efficiency, and other zero-emissions technology, including nuclear energy for those countries that choose to use it.

● Recognizing that a growing number of countries, including developing countries, have made considerable progress on their renewables and energy efficiency goals in line with their NDCs and carbon neutrality commitments.

⬤ Recognizing that this decade will be crucial for renewables and energy efficiency, with accelerated action and ambitious policy implementation that are vital to addressing energy security and affordability challenges.

⬤ The third pillar of the GDA will addresses methane and other non-CO_2 greenhouse gases through economy-wide methane-emission reduction. In support of this more than $1 billion will be mobilized for methane abatement projects, with additional information to be released on 5 December at the COP28 Energy Thematic Day.

⬤ The GDA also covers the Global Cooling Pledge, which targets substantially reducing global cooling emissions by 68 percent by 2050. Such emissions account for seven percent of the global total, a figure expected to triple as more nations adopt air-conditioning. As of today, 52 countries have signed the Pledge.

We declare our intent to work collaboratively and expeditiously to pursue the following objectives:

⬤ Commit to work together to triple the world's installed renewable energy generation capacity to at least 11,000 GW by 2030, taking into consideration different starting points and national circumstances.

⬤ Commit to work together in order to collectively double the global average annual rate of energy efficiency improvements from around 2% to over 4% every year until 2030.

⬤ Commit to put the principle of energy efficiency as the "first fuel" at the core of policymaking, planning, and major investment decisions

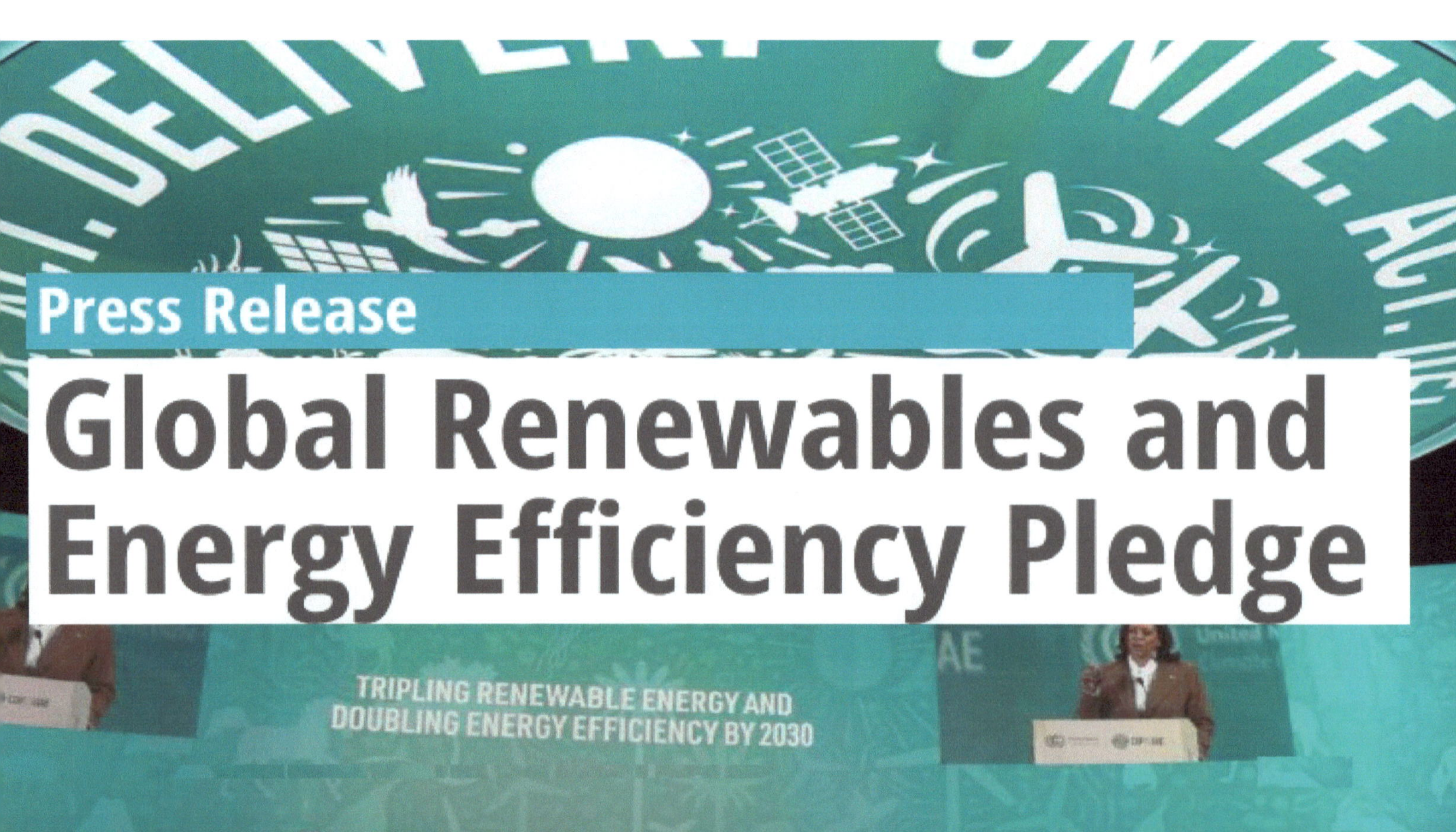

We declare our intent to work collaboratively and expeditiously to pursue the following objectives:

- Commit to take comprehensive domestic actions to contribute to the achievement of this pledge, including by adopting ambitious national policies on renewable energy and energy efficiency and reflecting this ambition in NDCs, working with cities and subnational governments, focusing on the key tools and enablers most relevant to national and local circumstances.

- Commit to ensure that policies are conducive to just energy transitions by empowering consumers and supporting the development of a skilled renewables and energy efficiency workforce, supporting current energy workers at risk of displacement by the energy transition, promoting productive reconversion of stranded assets, and ensure communities affected by this transition also benefit from the opportunities offered by the energy transition.

- Commit to ensuring that their efforts to scale up renewables and improve efficiency are conducted in an environmentally responsible manner.

- Commit to consider supporting existing international initiatives such as those outlined in the Power Breakthrough Agenda to advance technical and policy work that will serve to underpin actions under this pledge.

- Commit to agree on a way forward to review progress towards the Global Renewables and Energy Efficiency Pledge on an annual basis until 2030 for example by means of dedicated ministerial meetings and annual reports on the global progress towards targets of this pledge, marking use of existing flagship reports of the IEA and IRENA.

Source: https://www.ren21.net/wp-content/uploads/2023/12/COP-press-release.png
Reference link: https://www.cop28.com/en/global-renewables-and-energy-efficiency-pledge

DECLARATION ON CLIMATE AND HEALTH (COP28)

We, on the occasion of the first Health Day at the 28th UN Climate Change Conference (COP28), express our grave concern about the negative impacts of climate change on health. We stress the importance of addressing the interactions between climate change and human health and wellbeing in the context of the UNFCCC and the Paris Agreement, as the primary international, intergovernmental fora for the global response to climate change. We recognize the urgency of taking action on climate change, and note the benefits for health from deep, rapid, and sustained reductions in greenhouse gas emissions, including from just transitions, lower air pollution, active mobility, and shifts to sustainable healthy diets.

In order to work towards ensuring better health outcomes, including through the transformation of health systems to be climate-resilient, low-carbon, sustainable and equitable, and to better prepare communities and the most vulnerable populations for the impacts of climate change, we commit to pursuing the following common objectives:

● Strengthening the development and implementation of policies that maximize the health gains from mitigation and adaptation actions and prevent worsening health impacts from climate change, including through close partnerships with Indigenous Peoples, local communities, women and girls, children and youth, healthcare workers and practitioners, persons with disabilities and the populations most vulnerable to the health impacts of climate change, among others.

● Facilitating collaboration on human, animal, environment and climate health challenges, such as by implementing a One Health approach; addressing the environmental determinants of health; strengthening research on the linkages between environmental and climatic factors and antimicrobial resistance; and intensifying efforts for the early detection of zoonotic spill-overs as an effective means of pandemic prevention, preparedness and response.

● Recognizing that healthy populations contribute to, and are an effect of, climate resilience and an outcome of successful adaptation across a range of sectors - including food and agriculture, water and sanitation, housing, urban planning, health care, transport and energy - by prioritizing and implementing adaptation actions across sectors that deliver positive health outcomes.

● Improving the ability of health systems to anticipate, and implement adaptation interventions against, climate-sensitive disease and health risks, including by bolstering climate-health information services, surveillance, early warning and response systems and a climate-ready health workforce.

● Promoting a comprehensive response to address the impacts of climate change on health, including, for example, mental health and psychosocial wellbeing, loss of traditional medicinal knowledge, loss of livelihoods and culture, and climate-induced displacement and migration.

● Promoting a comprehensive response to address the impacts of climate change on health, including, for example, mental health and psychosocial wellbeing, loss of traditional medicinal knowledge, loss of livelihoods and culture, and climate-induced displacement and migration.

● Combating inequalities within and among countries, and pursuing policies that work towards accelerating achievement of the Sustainable Development Goals, including SDG3; reduce poverty and hunger; improve health and livelihoods; strengthen social protection systems, food security and improved nutrition, access to clean sources of energy, safe drinking water, and sanitation and hygiene for all; and work to achieve universal health coverage.

● Promoting steps to curb emissions and reduce waste in the health sector, such as by assessing the greenhouse gas emissions of health systems, and developing action plans, nationally determined decarbonization targets, and procurement standards for national health systems, including supply chains.

● Strengthening trans- and inter-disciplinary research, cross-sectoral collaboration, sharing of best practices, and monitoring of progress at the climate-health nexus, including through initiatives such as the Alliance for Transformative Action on Climate and Health (ATACH).

● Recognizing that health actors face challenges in accessing finance for health and climate change activities, particularly in low- and middle-income countries, we underscore the need to better leverage synergies at the intersection of climate change and health to improve the efficiency and effectiveness of finance flows.

● Encouraging the scaling up of investments in climate and health from domestic budgets, multilateral development banks, multilateral climate funds, health financing institutions, philanthropies, bilateral development agencies, and private sector actors.

● Encouraging international finance providers, including development banks, to strengthen the synergies between their climate and health portfolios, and enhance their support for country-led projects and programs in the health-climate nexus.

● Sharing learnings and best practices on financing and implementing climate-health interventions, and develop a common understanding of existing needs for climate-health finance, grounded in country priorities and needs. We welcome ongoing efforts in this regard, including by the COP28 presidency, the ATACH finance working group, and the joint Development Bank working group for climate-health financing.

● Improving monitoring, transparency and evaluation efforts of climate finance, as relevant, including for climate-health initiatives, in order to strengthen common understanding of its efficiency and effectiveness, and to maximize the delivery of positive health outcomes.

123 COUNTRIES ENDORSE

THE COP28 UAE DECLARATION ON CLIMATE AND HEALTH

To achieve these aims - according to our national circumstances - we commit to pursuing the better integration of health considerations into our climate policy processes, and of climate considerations across our health policy agendas, including by:

● Incorporating health considerations in the context of relevant Paris Agreement and UNFCCC processes, with a view to minimizing adverse effects on public health, and mainstreaming climate considerations in global health work programs, including those of the World Health Organization, where relevant and appropriate.

● Taking health into account, as appropriate, in designing the next round of nationally determined contributions, long term low greenhouse gas emission development strategies, national adaptation plans and adaptation communications.

THE GLOBAL METHANE PLEDGE

Methane is a powerful but short-lived climate pollutant that accounts for a third of net warming since the Industrial Revolution. Rapidly reducing methane emissions from energy, agriculture, and waste can achieve near-term gains in our efforts in this decade for decisive action and is regarded as the single most effective strategy to keep the goal of limiting warming to 1.5°C within reach while yielding co-benefits, including improving public health and agricultural productivity.

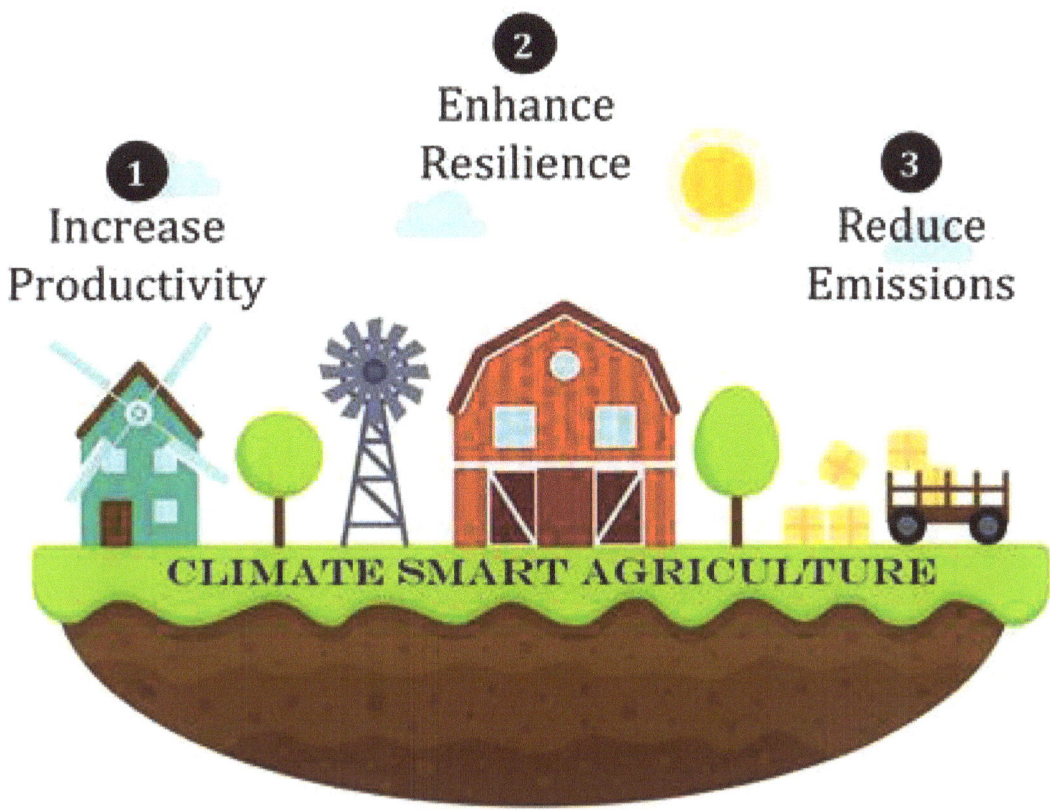

CLIMATE-SMART AGRICULTURE

(CLIMATE-RESILIENT AGRICULTURE/ WATER-SMART/ SOIL-SMART/ WEATHER-SMART/ ENERGY-SMART/ CARBON-SMART PRACTICES)

Climate-smart agriculture (CSA) is an integrated approach to managing landscapes—cropland, livestock, forests and fisheries--that address the interlinked challenges of food security and climate change. Climate change and food and nutrition insecurity pose two of the greatest development challenges of our time. Yet a more sustainable food system can not only heal the planet, but ensure food security for all.

Today, the global agrifood system emits one-third of all emissions. Global food demand is estimated to increase to feed a projected global population of 9.7 billion people by 2050. Traditionally, the increase in food production has been linked to agricultural expansion, and unsustainable use of land and resources. This creates a vicious circle, leading to an increase in emissions. Food systems are the leading source of methane emissions and biodiversity loss, and they use around 70% of fresh water. If food waste were a country, it would be the third highest emitter in the world. Meanwhile, emissions from agriculture are increasing in developing countries – a worrying trend which must be reversed.

Without significant climate mitigation action in the agri-food sector, the Paris Agreement goals cannot be reached. Agriculture is the primary cause of deforestation, threatening pristine ecosystems such as the Amazon and the Congo Basin. Without action, emissions from food systems will rise even further, with increasing food production.

CLIMATE-RESILIENT AGRICULTURE

Climate-resilient agriculture (CRA) is an approach that includes sustainably using existing natural resources through crop and livestock production systems to achieve long-term higher productivity and farm incomes under climate variabilities.

This practice reduces hunger and poverty in the face of climate change for forthcoming generations. CRA practices can alter the current situation and sustain agricultural production from the local to the global level, especially in a sustainable manner.

Improved access and utilization of technology, transparent trade regimes, increased use of resources conservation technologies, an increased adaptation of crops and livestock to climatic stress are the outcomes from climate-resilient practices.

WATER-SMART AGRICULTURE

WaSA is an approach to efficiently harvesting, storing, and channeling green water throughout the year, regardless of season. Rather than hoping for rain to water fields, rural poor farmers with constrained resources can harvest rain when it comes and use less water year-round. This alone can double incomes for farmers, who no longer have to rely solely on incomes from rainy season farming but can plant, grow, and harvest during the dry season as well.

WaSA is not a new concept. It draws from water-related interventions of well-known approaches, particularly Climate Smart Agriculture, sustainable agriculture, and conservation agriculture. However, WaSA draws attention to access to water for production, including increasing the soil's capacity to absorb and store moisture (green water), rainwater harvesting and storage, wastewater reuse, and supplementary small-scale irrigation. In doing so, WaSA emphasizes collaboration across the water and agriculture sectors in order to ensure increased investment in smallholder agriculture and efficient use of water for production, with a more pointed goal of increasing water access for production among smallholder farmers, and increasing yields, incomes, food security, and nutrition.

SOIL SMART AGRICULTURE

Focus:
Optimizing soil health and fertility.
Practices:
Cover cropping: Planting crops between main crops to improve soil structure, nutrient cycling, and erosion control.
No-till farming: Minimizing soil disturbance to preserve organic matter and reduce erosion.
Precision agriculture:
Using technology to apply inputs (e.g., fertilizers, pesticides) based on specific soil needs.

Focus: Adapting to changing weather patterns and climate variability.

Practices:

Weather forecasting: Using advanced weather data to make informed decisions about planting, harvesting, and irrigation.

Drought-tolerant varieties: Cultivating crops that can withstand dry conditions.

Irrigation efficiency: Implementing efficient irrigation systems to conserve water.

ENERGY SMART AGRICULTURE

Focus: Reducing energy consumption and greenhouse gas emissions.

Practices:

Renewable energy: Utilizing solar, wind, or biomass energy for agricultural operations.

Energy-efficient equipment: Using machinery and equipment that consume less energy.

Integrated pest management (IPM): Employing natural pest control methods to reduce the need for chemical pesticides.

CARBON SMART AGRICULTURE

Focus:
Sequestering carbon in the soil to mitigate climate change.
Practices:
Soil carbon sequestration: Implementing practices like cover cropping and no-till farming to increase organic matter in the soil.
Agroforestry:
Integrating trees with agricultural crops to enhance carbon storage and biodiversity.
Reduced tillage:
Minimizing soil disturbance to preserve soil carbon.

CAN 'NATURAL GAS' BE CONSIDERED AS ' CLEAN ENERGY OR GREEN ENERGY' SOURCE?

PROS:

Lower greenhouse gas emissions: Compared to coal and oil, natural gas produces fewer greenhouse gases like carbon dioxide.

Energy efficiency: Natural gas is a relatively efficient fuel source.

Infrastructure: It can be easily integrated into existing energy infrastructure.

CONS:

Methane emissions: Methane, a potent greenhouse gas, can leak during natural gas production, transportation, and distribution.

Fossil fuel: Natural gas is a fossil fuel, meaning it's a finite resource.

Environmental impacts: Natural gas extraction can have environmental consequences, such as groundwater contamination and habitat destruction.

Source: https://th.bing.com/th/id/OIP.6iTk-i8-tLO-QitiZ_GUDwHaE8?rs=1&pid=ImgDetMain
Reference link: https://gemini.google.com/

ENVIRONMENTAL PERFORMANCE INDEX

The Environmental Performance Index (EPI) is a method of quantifying and numerically marking the environmental performance of a state's policies. This index was developed from the Pilot Environmental Performance Index, first published in 2002, and designed to supplement the environmental targets set forth in the United Nations Millennium Development Goals. The EPI was preceded by the Environmental Sustainability Index (ESI), published between 1999 and 2005. Both indices were developed by Yale University (Yale Center for Environmental Law and Policy) and Columbia University (Center for International Earth Science Information Network) in collaboration with the World Economic Forum and the Joint Research Centre of the European Commission. The ESI was developed to evaluate environmental sustainability relative to the paths of other countries. Due to a shift in focus by the teams developing the ESI, the EPI uses outcome-oriented indicators, then working as a benchmark index that can be more easily used by policy makers, environmental scientists, advocates and the general public. Other leading indices like the Global Green Economy Index (GGEI) provide an integrated measure of the environmental, social and economic dynamics of national economies. The GGEI utilizes EPI data for the environmental dimension of the index while also providing a performance assessment of efficiency sectors (e.g. transport, buildings, energy), investment, green innovation and national leadership around climate change. The EPI for the year 2022 ranks 180 countries. The top five countries are Denmark, United Kingdom, Finland, Malta and Sweden. India ranked last at 180 with a score of 18.9. The 2024 Environmental Performance Index (EPI) provides a data-driven summary of the state of sustainability around the world. Using 58 performance indicators across 11 issue categories, the EPI ranks 180 countries on climate change performance, environmental health, and ecosystem vitality. These indicators provide a gauge at a national scale of how close countries are to established environmental policy targets. The EPI offers a scorecard that highlights leaders and laggards in environmental performance and provides practical guidance for countries that aspire to move toward a sustainable future. EPI indicators provide a way to spot problems, set targets, track trends, understand outcomes, and identify best policy practices. Going beyond the aggregate scores and drilling down into the data to analyze performance by issue category, policy objective, peer group, and country offers even greater value for policymakers. This granular view and comparative perspective can assist in understanding the determinants of environmental progress and in refining policy choices.

Source: https://epi.yale.edu/sites/default/files/images/EPI_Logo_AI_Master_file.svg
Reference link: https://en.m.wikipedia.org/wiki/Environmental_Performance_Index, https://epi.yale.edu/

FIRST EARTH SUMMIT, 1992, RIO

The United Nations Conference on Environment and Development (UNCED), also known as the 'Earth Summit', was held in Rio de Janeiro, Brazil, from 3-14 June 1992. This global conference, held on the occasion of the 20th anniversary of the first Human Environment Conference in Stockholm, Sweden, in 1972, brought together political leaders, diplomats, scientists, representatives of the media and non-governmental organizations (NGOs) from 179 countries for a massive effort to focus on the impact of human socio-economic activities on the environment.

A 'Global Forum' of NGOs was also held in Rio de Janeiro at the same time, bringing together an unprecedented number of NGO representatives, who presented their own vision of the world's future in relation to the environment and socio-economic development.The Rio de Janeiro conference highlighted how different social, economic and environmental factors are interdependent and evolve together, and how success in one sector requires action in other sectors to be sustained over time. The primary objective of the Rio 'Earth Summit' was to produce a broad agenda and a new blueprint for international action on environmental and development issues that would help guide international cooperation and development policy in the twenty-first century.The 'Earth Summit' concluded that the concept of sustainable development was an attainable goal for all the people of the world, regardless of whether they were at the local, national, regional or international level. It also recognized that integrating and balancing economic, social and environmental concerns in meeting our needs is vital for sustaining human life on the planet and that such an integrated approach is possible. The conference also recognized that integrating and balancing economic, social and environmental dimensions required new perceptions of the way we produce and consume, the way we live and work, and the way we make decisions. This concept was revolutionary for its time, and it sparked a lively debate within governments and between governments and their citizens on how to ensure sustainability for development.

One of the major results of the UNCED Conference was Agenda 21, a daring program of action calling for new strategies to invest in the future to achieve overall sustainable development in the 21st century. Its recommendations ranged from new methods of education, to new ways of preserving natural resources and new ways of participating in a sustainable economy. The 'Earth Summit' had many great achievements: the Rio Declaration and its 27 universal principles, the United Nations Framework Convention on Climate Change (UNFCCC), the Convention on Biological Diversity; and the Declaration on the principles of forest management. The 'Earth Summit' also led to the creation of the Commission on Sustainable Development, the holding of first world conference on the sustainable development of small island developing States in 1994, and negotiations for the establishment of the agreement on straddling stocks and highly migratory fish stocks.

CENTRE FOR CLIMATE REPORTING

● The Centre for Climate Reporting is a not-for-profit investigative journalism organization focused on the biggest story of our time: climate change.

● Multinational corporations, high net worth individuals, populist movements, petrostates – across the world, powerful forces are obstructing efforts to tackle climate change and threaten even democracy itself.

● Through bold, cross-border investigative reporting, our mission is to uncover how these actors build and wield political power – to hold them to account.

● To ensure that these stories reach as wide an audience as possible, we partner and collaborate with the world's most respected media organisations.

Cheap cars, supersonic jets and floating power plants: Undercover in Saudi Arabia's secretive program to keep the world burning oil. Speaking to undercover reporters, Saudi energy officials disclosed ambitious plans to undo progress on phasing out oil by financing high carbon infrastructure across Africa and Asia.

● We report without fear or favour and are committed to rigorous, fair, public interest journalism.
● The Centre for Climate Reporting is strictly editorially independent.
● We are committed to working with integrity and fairness to uncover wrongdoing.

CLIMATE ACTION AGAINST DISINFORMATION (CAAD)

CAAD is dedicated to combatting climate mis/disinformation, uniting policy, research, and accountability initiatives to stimulate a meaningful global dialogue on effective climate solutions. They are steadfast in their commitment to breaking down one of the most significant obstacles to tackling climate change: the spread of deceptive and false narratives that undermine constructive discussions about our environment.CAAD spearheads efforts in policy development, communication strategies, and research to hold Big Tech responsible for their part in facilitating the circulation of climate misinformation on their platforms.

Moreover, CAAD actively collaborates with decision-makers at both national and international levels to foster political awareness of this critical issue and to reinforce accountability measures.

From exposing platforms' failure to curb misinformation to establishing COP Intelligence Units to monitor the most harmful false narratives circulated during the annual global climate conference, CAAD tirelessly advocates for accountability. Their goal is to foster a transparent and honest dialogue on environmental matters, ultimately paving the way for substantial and impactful climate solutions.

CLIMATE MONITORING AGENCIES

A climate change monitoring system integrates satellite observations, ground-based data and forecast models to monitor and forecast changes in the weather and climate. A historical record of spot measurements is built up over time, which provides the data to enable statistical analysis and the identification of mean values, trends and variations. The better the information available, the more the climate can be understood and the more accurately future conditions can be assessed, at the local, regional, national and global level. This has become particularly important in the context of climate change, as climate variability increases and historical patterns shift.

DESCRIPTION OF CLIMATE CHANGE MONITORING AND EVALUATION

Systematic observation of the climate system is usually carried out by national meteorological centres and other specialised bodies. They take measurements and make observations at standard preset times and places, monitoring atmosphere, ocean and terrestrial systems. Since national monitoring systems all form part of a global network, it is vital that there is as much consistency as possible in the way measurements and observations are made. This includes accuracy, the variables measured and the units they are measured in. The World Meteorological Organisation (WMO) performs a vital role in this respect. The National Meteorological or Hydrometeorological Services (NMHS) of 189 member states and territories form the membership of the WMO. This enables the WMO to establish and promote best practice in national climate monitoring, provide support to the NMHSs and effectively implement specific initiatives.

In 1992 the Global Climate Observing System (GCOS) was established to ensure that the observations and information needed to address climate-related issues are obtained and made available to all potential users. The initiative was co-sponsored by the WMO, the Intergovernmental Oceanographic Commission (IOC) of UNSECO, the United Nations Environment Programme (UNEP) and the International Council for Science (ICSU). The stated goal of GCOS is: "to provide comprehensive information on the total climate system, involving a multidisciplinary range of physical, chemical and biological properties, and atmospheric, oceanic, hydrological, cryospheric and terrestrial processes. GCOS is intended to meet the full range of national and international requirements for climate and climate-related observations. As a system of climate-relevant observing systems, it constitutes, in aggregate, the climate observing component of the Global Earth Observation System of Systems (GEOSS)".

As part of its role to provide vital and continuous support to the United Nations Framework Convention on Climate Change (UNFCCC), GCOS has established 20 Climate Monitoring Principles, as well as defining 50 Essential Climate Variables (ECVs).

RESPONSIBLE BUSINESS

The "Quintuple Bottom Line": Profit, People, Planet, Ethics, and Equity.Responsible Business is the practice of creating customer value through the active concern for people, ethics, equity, and environmental impacts while running a profitable business.

The Quintuple Bottom Line (QBL) of Responsible Business enables us to view business practices and challenges through the lens of Profit, People, Planet, Ethics, and Equity. By focusing on Profit, a business ensures it is financially sustainable. Without generating a profit, a business cannot survive or thrive, so the primary role of a responsible business owner is to ensure that the business returns a profit and provides income to survive and grow.

Unless it makes a profit, the business cannot fulfil the other four aspects of the QBL.By paying attention to People, a business focuses on its employees, suppliers, and customers. Employees are the key group that make the business function and enable it to create value-added. Satisfied, happy, and productive employees are the lifeblood of a successful business. They deserve good wages, meaningful employment, and the means to sustain their families and live meaningful lives. Suppliers are key because they enable the business to obtain inputs that are goal-effective and cost-effective in order to generate value. Customers create the demand for its products and deserve safe, high quality, and high value products and services.

Planet refers to the responsibility of the business to operate in ways that minimize the harm to the environment and the planet, through its actions. It treats the natural environment as a finite resource, which must be sustained and respected. Environmental degradation and destruction lead to a situation in which business itself is no longer sustainable. Ethics is a code by which a business abides, which includes legal, moral, and values-based considerations that guide business behaviour. Ethics includes operating with integrity and in a transparent manner; that is, "doing it right" with nothing to hide.

"Do right to do good" is the guiding principle of responsible business.Equity is addressed within the context of community. Equity requires that a responsible business acts in a way that provides for the health of the community in which it operates, providing local employment for its citizens, providing programs that help the needy and the underrepresented, and, in general, building a sense of equitable community. These actions reflect the responsibility of contributing to the improvement of the quality of life in the communities in which it operates and for acting in a manner that ensures the equity of the process to everyone it touches.

CIRCULAR ECONOMY (ISO 59000)

ISO 59004 is a part of the ISO 59000 family of standards, specifically designed to foster a shift towards a circular economy. This standard provides comprehensive guidance applicable to any type of organization. It includes defining key terms and concepts, outlining a vision for a circular economy, elucidating core principles, and offering practical guidance for actionable steps towards sustainability. The standard aims to support organizations in contributing to the United Nations Agenda 2030 for Sustainable Development by facilitating a transition to a circular use of resources.

WHY IS ISO 59004 IMPORTANT?

The linear model of the global economy—characterized by extraction, production, use, and disposal—has led to significant environmental challenges, including resource depletion climate change and biodiversity loss. ISO 59004 addresses these issues by advocating for a circular economy model, which emphasizes the sustainable management and renewal of natural resources. By adopting this standard, organizations can:

- Deliver more sustainable and ambitious solutions.
- Improve relationships with stakeholders.
- Engage in more effective and efficient fulfilment of voluntary and legal obligations.
- Contribute to climate change mitigation and adaptation.
- Increase resilience against resource scarcity and other environmental, social, and economic risks.

Adherence to ISO 59004 enables organizations to create and share more value within society while ensuring the quality and resilience of ecosystems, ultimately supporting a sustainable future.

Source: https://africaninsightacademy.co.za/embracing-the-circular-economy-a-paradigm-shift-towards-sustainable-living/ Reference link: https://www.iso.org/standard/80648.html

ENERGY EFFICIENCY

Energy Efficiency means using less energy to get the same job done – and in the process, cutting energy bills and reducing pollution. Many products, homes, and buildings use more energy than they actually need, through inefficiencies and energy waste. Energy efficiency is one of the easiest ways to eliminate energy waste and lower energy costs. It is also one of the most cost-effective ways to combat climate change, clean the air we breathe, help families meet their budgets, and help businesses improve their bottom lines. Millions of American consumers and businesses choose or invest in energy-efficient products.

Examples of energy efficiency

Anywhere that energy is used; there is an opportunity to improve efficiency. Some products, such as your HVAC system and water heater, are major energy-users and can experience huge energy and cost savings through improvements in energy efficiency. Additionally, other products don't use energy directly, but they improve the overall efficiency and comfort of a house or a building (such as thermal insulation or windows).

Heat pumps: Heat pumps are an efficient way to heat and cool your home because they move heat from the surrounding air, instead of creating it. Heat pumps can also do double-duty and both heat and cool your home year-round, as opposed to relying on two separate systems to condition your space.

Heat pump water heaters: Your water heater if the second highest energy user in your home. On average, an ENERGY STAR certified heat pump water heater uses 70% less energy and can help a family of four save over $550 a year compared to a standard electric water heater.
Windows: Energy-efficient windows are made with materials that reduce heat exchange and an air leak, which means you don't need as much energy to heat or cool a space.

Insulation: Adding more insulation to an attic keeps the warm air inside from escaping in the winter. In the summer, it keeps hot air out. With good insulation, you won't need to use as much energy to keep your house warm in the winter or cool in the summer.

Smart thermostats: Smart thermostats are Wi-Fi enabled devices that control heating and cooling in your home by learning your temperature preferences and schedule to automatically adjust to energy-saving temperatures when you are asleep or away. They can help you lower your energy bills by not spending money to heat or cool an empty house.

Computer power management: Computers can be set to automatically enter a low-power "sleep" mode when not in use.

Source: https://ushomefilter.com/blogs/air-quality/energy-efficiency-meaning Reference link: https://www.energystar.gov/about/how-energy-star-protects-environment/energy-efficiency#:~:text=Simply%20put%2C%20energy%20efficiency%20means,through%20inefficiencies%20and%20energy%20waste.

NETWORK FOR GREENING THE FINANCIAL SYSTEM

"Networking for Greening the Financial System" (NGFS) is a global network of central banks and financial supervisors that aims to enhance the role of the financial system in managing climate and environmental risks. The network was established in 2017 during the "One Planet Summit" in Paris. Its primary goals include:

SHARING BEST PRACTICES: NGFS members collaborate to share insights, tools, and strategies that can be adopted to address climate-related financial risks.

ADVANCING CLIMATE FINANCE: The network works to integrate climate-related risks into the financial sector, encouraging sustainable investments and promoting green finance.

DEVELOPING SUPERVISORY TOOLS: NGFS develops and promotes the use of supervisory tools and methods to assess and manage climate-related risks.

PROVIDING TECHNICAL ASSISTANCE: The network supports central banks and supervisors, particularly in emerging markets, by providing technical assistance on greening the financial system.

PROMOTING SUSTAINABLE INVESTMENTS: By encouraging the incorporation of climate-related risks into financial decisions, NGFS aims to mobilize the financial sector to support the transition to a sustainable economy.

NGFS has released several reports and guidelines that provide recommendations for central banks, regulators, and financial institutions on integrating climate-related risks into their operations and strategies. This network plays a crucial role in the global effort to align the financial system with the goals of the Paris Agreement and broader sustainability objectives.

IFRS S1 AND S2

IFRS S1: General Requirements for Disclosure of Sustainability-related Financial Information. The main objective of this Standard is to disclose all information about sustainability-related risks and opportunities that could reasonably be expected to affect a company's prospects. IFRS S1 provides the basic requirements for sustainability disclosures, which should be used with IFRS S2 as well as the future Standards the ISSB releases. The Standardrequires disclosure of material information about sustainability-related risks and opportunities with the financial statements, to meet investor information needs-requires industry specific disclosures and refers to the industry-based SASB standards for guidance when identifying disclosures about sustainability-related risks or opportunities-refers to sources to help companies identify sustainability-related risks and opportunities and information (for everything other than in the scope of IFRS S2 requires disclosures that enable investors to understand the connections between the sustainability-related risks and opportunities and the sustainability-related financial disclosures and financial statementsis GAAP agnostic.

Source: https://quaderno.io/blog/how-to-comply-with-ifrs-accounting-standards/ Reference link: https://www.grantthornton.global/en/insights/articles/overview-of-ifrs-s1-and-ifrs-s2/

IFRS S2: CLIMATE-RELATED DISCLOSURES

The two Standards are designed to be applied together. However, IFRS S2 has been developed to capture a climate-specific requirement which includes:Strategy disclosures that distinguish between physical and transitional risks. Disclosure of their plans to respond to climate-related risks and opportunities, including how climate-related targets are set and any targets it is required to meet by law or regulation. Companies should perform scenario analysis to explain how various climate-related events may impact the business in the future. Climate-related metrics and target disclosures should include:Cross-industry metrics that are relevant to all companies e.g. greenhouse gas emissions, refer to our publication on 'What are sustainability scope 1, 2 and 3 emissions?' for more information on greenhouse gas emissions.Industry-based metrics relevant to companies within the related industries andcompany specific metrics considered by the board or management when measuring progress towards set targets.

ENERGY CONSERVATION

Energy conservation is the effort to reduce wasteful energy consumption by using fewer energy services. This can be done by using energy more effectively (using less energy for continuous service) or changing one's behaviour to use less service (for example, by driving less). Energy conservation can be achieved through efficient energy use, which has some advantages, including a reduction in greenhouse gas emissions and a smaller carbon footprint, as well as cost, water, and energy savings. Energy can be conserved by reducing waste and losses, improving efficiency through technological upgrades, improving operations and maintenance, changing users' behaviours through user profiling or user activities, monitoring appliances, shifting load to off-peak hours, and providing energy-saving recommendations. Observing appliance usage, establishing an energy usage profile, and revealing energy consumption patterns in circumstances where energy is used poorly, can pinpoint user habits and behaviors in energy consumption. Appliance energy profiling helps identify inefficient appliances with high energy consumption and energy load. Seasonal variations also greatly influence energy load, as more air-conditioning is used in warmer seasons and heating in colder seasons.

Energy efficiency and energy conservation are related and often complimentary or overlapping ways to avoid or reduce energy consumption. Energy efficiency generally pertains to the technical performance of energy conversion and energy-consuming devices and to building materials. Energy conservation generally includes actions to reduce the amount of end-use energy consumption.

Efficiency and conservation measures can help to directly lower consumers' energy bills and potentially reduce greenhouse gas emissions associated with energy use. Consumers also benefit indirectly when reducing their electricity consumption helps to reduce demand on the electric system. High electricity demand often results in higher costs for generating and transmitting electricity that may be passed on to utility customers.

Examples of energy efficiency and conservation measures for consumers include:

- Buying energy-efficient products and vehicles with high fuel economy
- Using programmable thermostats to control heating and cooling systems
- Installing energy management and control systems in commercial and industrial facilities
- Turning off lights and electric appliances when not in use
- Participating in energy efficiency and conservation programs that utilities offer their customers

Source: https://www.agwayenergy.com/blog/7-ways-to-celebrate-national-energy-conservation-day/ Reference link: https://en.wikipedia.org/wiki/Energy_conservationhttps://www.eia.gov/energyexplained/use-of-energy/efficiency-and conservation.php#:~:text=Energy%20conservation%20generally%20includes%20actions,switches%2C%20is%20a%20conservation%20measure.

ENERGY TRANSITION

In 2023, according to the United Nations World Meteorological Organization, the global annual average temperature was nearly 1.5 degrees Celsius above pre-industrial levels (more precisely, 1.45 ± 0.12 °C). This is a dramatic figure, because the 2015 Paris Agreement on climate change aims to limit the long-term temperature rise (i.e., the average for a decade, rather than a single year such as 2023) to no more than 1.5 degrees Celsius. In addition to causing the polar ice caps to melt and sea levels to rise, global warming is causing other types of climate change, like desertification and an increase in extreme weather events such as hurricanes, floods and fires: the distortion of the climate risks causing incalculable damage. The scientific community is in agreement that this is due to anthropogenic emissions of greenhouse gases into the atmosphere, especially since the Industrial Revolution. The main such gas, carbon dioxide, originates largely from the energy sector (including but not limited to the generation of electricity).

In December 2023, COP28 in Dubai closed with an explicit agreement to put an end to the use of fossil fuels, but it did not set precise targets to phase out nonrenewable energy sources. At the same time, it admitted that the world's countries are not yet on track to meet the goal of containing the global temperature rise to within 1.5° C. COP28 called on parties to take action on a global scale to triple renewable energy capacity and double progress in energy efficiency by 2030. It also called on them to put forward ambitious emission reduction targets, covering all greenhouse gases (GHGs), economic sectors and categories. This is in line with the 1.5°C limit in the next round of national climate action plans for 2025. The goal by 2050 is still to achieve so-called Carbon Neutrality, in other words, to reduce and avoid greenhouse emissions by offsetting the remaining emissions through the use of so-called carbon credits.

Source: https://www.tno.nl/en/about-tno/organisation/units/energy-transition/ Source: https://en.m.wikipedia.org/wiki/Energy_transition Reference link: https://www.enelgreenpower.com/learning-hub/energy-transition

To achieve this goal, which was enshrined at COP26 in Glasgow, our main tool is the energy transition, i.e., the shift from an energy mix based on fossil fuels to one that produces very limited, if not zero, carbon emissions, based on renewable energy sources. A huge contribution to decarbonization comes from the electrification of consumption, replacing fossil fuel-generated electricity with energy generated from renewable sources, which also makes other sectors like transport cleaner; the digitalization of grids also contributes by improving energy efficiency. Historically speaking, energy transitions are not new. In the past we have seen huge epoch-marking shifts like the transition from using wood to using coal in the 19th century or from coal to oil in the 20th century. But what distinguishes this transition from its predecessors is the urgency of protecting the planet from the greatest threat it has ever had to face, and of doing so as quickly as possible. This impetus has accelerated the changes in the energy sector: between 2010 and 2022, according to Irena data, the cost of renewable technologies decreased by 83% in the case of solar PV and 42% in the case of onshore wind. The energy transition is not, however, only limited to the gradual closure of coal-fired power stations and the development of clean energies: it is a paradigm shift that concerns the entire system. This solution can provide benefits not only for the climate but also for the economy and for society. The digitalization of electricity grids can usher in the age of smart grids and open the way for new services for consumers. From the environmental perspective, renewable sources and electric mobility reduce pollution, while coal-fired power stations can be repurposed in line with the principles of the circular economy.

Concerning social sustainability, the new jobs created can absorb those people who were previously working in the thermoelectric sector. It is important that the energy transition be inclusive and ensure that no one is left behind.

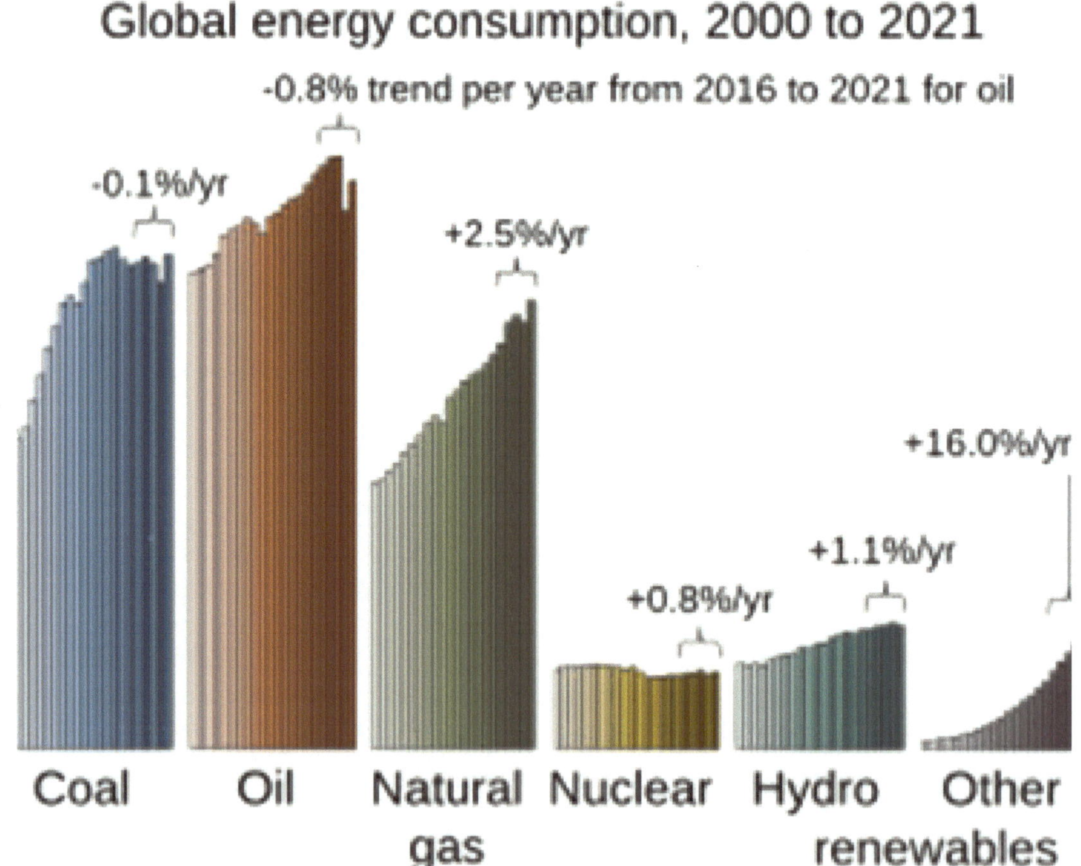

AFFORESTATION
REFORESTATION

Afforestation and reforestation are both methods of planting trees to create forests:

Afforestation

Planting trees or sowing seeds in an area that was previously unforested. Afforestation can help promote biodiversity by incorporating native plant species and providing habitats for wildlife.

Reforestation

Replanting trees in an area that was previously forested but has been damaged or destroyed by natural disasters, logging, or other human activities. Reforestation aims to restore the original forest cover and help the ecosystem recover.

Afforestation (i.e. converting long-time non-forested land into forest) refers to the establishment of forests where previously there have been none, or where forests have been missing for a long time (50 years according to UNFCCC) while reforestation refers to the replanting of trees on more recently deforested land (i.e. converting recently non-forested land in forest). If these two approaches are viewed as complementary, they may enable "win-win" policy options. However, if unsustainably managed, both practices may be controversial as they may lead to the destruction of original non-forest ecosystems (e.g. natural grassland).

Reforestation and afforestation are two of the leading nature-based solutions for tackling the effects of climate change. For commercial foresters and landowners, these two practices are essential to ensuring they can grow wood for wood products and continuously meet demand in a sustainable way. Reforestation is crucial in combating or preventing deforestation or forest degradation, where forests shrink in size or are completely removed. As well as reducing a forest's ability to absorb carbon dioxide (CO_2), deforestation can destroy wildlife habitats and contribute to the likelihood of flooding in certain areas. Afforestation can also help avoid desertification, where fertile land turns into a desert as a result of drought or intensive agriculture.

Reference link: https://climate-adapt.eea.europa.eu/en/metadata/adaptation-options/afforestation-and-reforestation-as-adaptation-opportunity#:~:text=Afforestation%20(i.e.%20converting%20long%2Dtime,on%20 more%20recently%20deforested%20land%20(

PALEONTHOLOGIST

A paleontologist is a scientist who studies the history of life on Earth through the fossil record. Fossils are the evidence of past life on the planet and can include those formed from animal bodies or their imprints (body fossils). Trace fossils are another kind of fossil. A trace fossil is any evidence of the life activity of an animal that lived in the past. Burrows, tracks, trails, feeding marks, and resting marks are all examples of trace fossils. A paleontologist is a scientist who studies the history of life on Earth by analysing fossils. Fossils can include evidence of past animal life, such as body fossils and trace fossils, which are signs of animal activity like tracks, burrows, or feeding marks. Paleontologists use the information they uncover from fossils to learn about past life and the Earth's environment.

THEIR WORK CAN INCLUDE:

Research: Studying specimens in museums, finding new fossils, analysing rocks, and more
Excavation: Attending digs to uncover fossils and other samples
Lab work: Reviewing collected items and performing tests to identify markers like age and pecies Paleontologists often specialize in a particular era or type of fossil. Their research can help them understand the relationships between extinct animals and plants and their living relatives, reconstruct ancient communities, and study extinction events. This information can also help paleontologists predict how climate change might affect organisms in the future.

PALEOCLIMATOLOGIST

Paleoclimatologists study Earth's climate history by analysing environmental evidence to understand past conditions and why they occurred. They use proxy climate records, or natural recorders of environmental change, to estimate past conditions and extend our understanding of climate back hundreds to millions of years. These records include:

- Tree rings: Also known as dendrochronology, this method helps paleoclimatologists under stand the general temperature or precipitation that occurred when the tree was alive
- Shells of deceased marine creatures
- Ice cores
- Glaciers and ice caps
- Laminated sediments from lakes and the ocean

Paleoclimatology has taught scientists that there have been several ice ages and periods of global warming, and that there can be both short- and long-term trends in climate. Paleoclimatologists can work in a variety of settings, including:

- Federal government: 26% of paleoclimatologists work in this sector
- Education system: 20% of paleoclimatologists work in this sector, typically in teaching or support roles
- Television broadcasting: 6% of paleoclimatologists work in this sector, either researching documentaries or consulting on fiction and factual production

Paleoclimatology is the study of previous climates that have existed during Earth's different geologic ages.Since modern records do not outline most of Earth's climatic past, scientists must gather data preserved in nature over the millennia in paleological remains referred to as proxy records.

Reference link: https://education.nationalgeographic.org/resource/paleoclimatology-RL/ https://bifrostonline.org/wp-content/uploads/2018/10/Measuring_climate_change_without_thermometers.jpg

PALEOCEANOGRAPHERS

The environmental information is combined with that on planktic and benthic microfossils, so that interactions can be studied between fluctuations in oceanic environments and oceanic biota. Because paleoceanography uses properties of components of oceanic sediments, it is limited in its scope and time-resolution by the sedimentary record.

Paleoceanography offers a view of a world alternative to, and different from, our present world, including colder ('ice ages') as well as warmer worlds, providing data on boundary conditions of an ocean–atmosphere system very different from those in the present world. In addition, paleoceanography provides information on rates of climate change in the past, and possible linkages (or lack thereof) between climate change and atmospheric pCO_2 levels. We use paleoceanographic data to evaluate in which aspects the present world can be a guide to the past, in which aspects the ocean–atmosphere system of the present world with its present biota is only one possible, not the only, stable mode of the Earth system. Paleoceanography thus enables us to use the past in order to gain information on possible future climatic and biotic developments.

Paleoceanography is the study of the history of the oceans in the geologic past with regard to circulation, chemistry, biology, geology and patterns of sedimentation and biological productivity. Paleoceanographic studies using environment models and different proxies enable the scientific community to assess the role of the oceanic processes in the global climate by the re-construction of past climate at various intervals. Paleoceanographic research is also intimately tied to paleoclimatology.

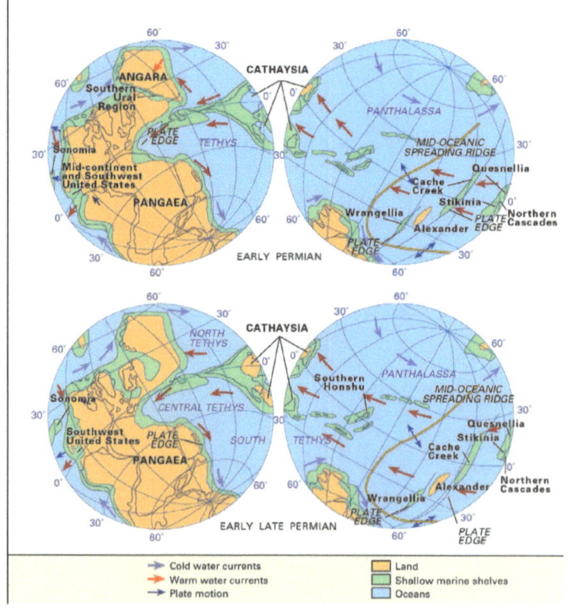

Reference link: https://www.sciencedirect.com/topics/earth-and-planetary-sciences/paleoceanography#:~:text=Paleoceanography%20is%20the%20study%20of,physical%2C%20chemical%20and%20biological%20character. https://en.wikipedia.org/wiki/Paleoceanography https://cdn.britannica.com/03/803-050-697CFBD3/Paleogeography-paleoceanography-times-Early-Permian.jpg

ECO-SYSTEMS

An ecosystem is a community of living organisms and their physical environment that interact with each other. Ecosystems can be natural or artificial, land-based or water-based, and can be of any size. They include living parts, such as animals, plants, fungi, and microscopic organisms, and non-living parts, such as the sun, water, air, and rocks. These parts are linked together through nutrient cycles and energy flows, and are affected by internal and external factors.

Some examples of ecosystems include:

Ecosystems are the foundation of the Biosphere and determine the health of the entire Earth system. The healthier our ecosystems are, the healthier the planet and its people. Human activities, such as polluting, clearing land, and introducing invasive species, can disturb ecosystems. The UN Decade on Ecosystem Restoration aims to prevent, halt, and reverse the degradation of ecosystems on every continent and in every ocean.

 Environment involves both living organisms and the non-living physical conditions. These two are inseparable but inter-related. The living and physical components are linked together through nutrient cycles and energy flows.

Reference link: https://australian.museum/learn/species-identification/ask-an-expert/what-is-an-ecosystem/#:~:text=An%20ecosystem%20is%20a%20community%20of%20organisms%20and%20their%20physical,nutrient%20cycles%20and%20energy%20flows.

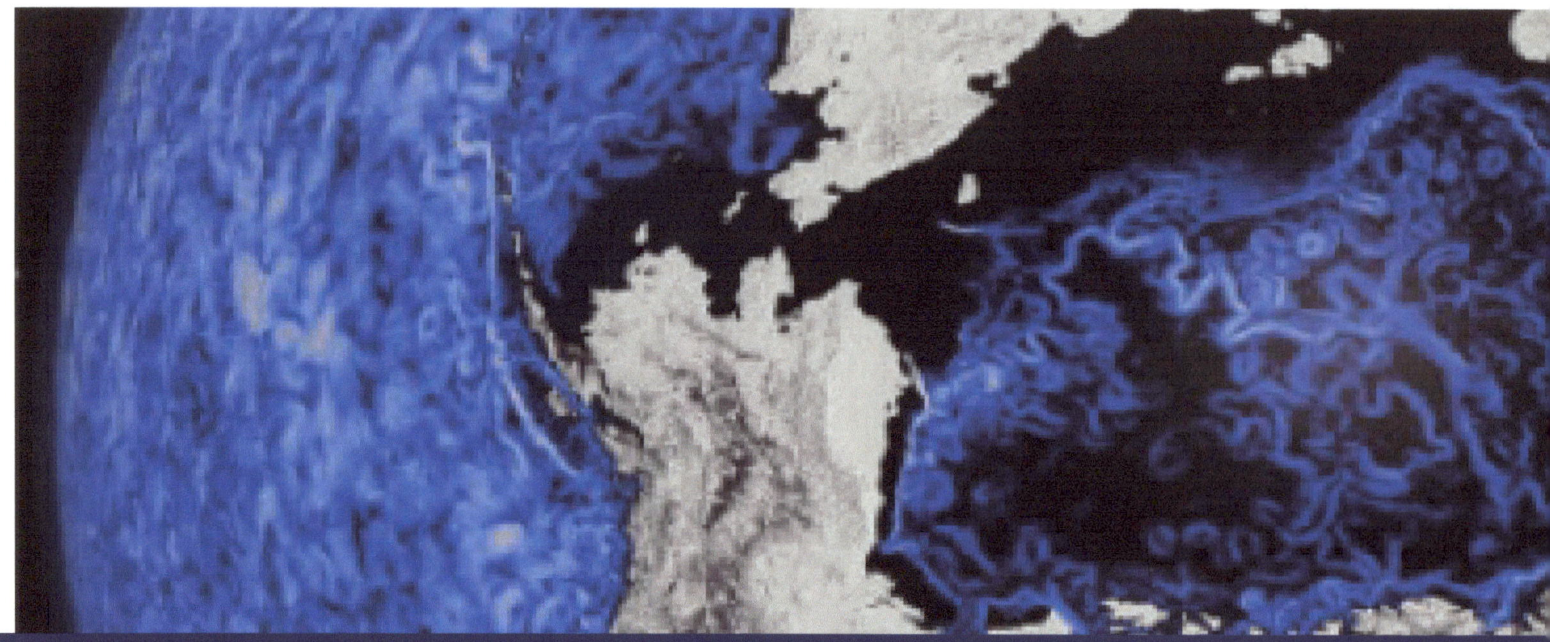

EARTH SYSTEM

What is the most important part of our planet, the main reason Earth is different from all the other planets in the solar system? If 10 different environmental scientists were asked this question, they would probably give 10 different answers. Each scientist might start with their favourite topic, from plate tectonics to rainforests and beyond. Eventually, however, their collective description would probably touch on all the major features and systems of our home planet. It turns out that no single feature is more significant than the others—each one plays a vital role in the function and sustainability of Earth's system.

There are five main systems, or spheres, on Earth. The first system, the geosphere, consists of the interior and surface of Earth, both of which are made up of rocks. The limited part of the planet that can support living things comprises the second system; these regions are referred to as the biosphere. In the third system are the areas of Earth that are covered with enormous amounts of water, called the hydrosphere. The atmosphere is the fourth system, and it is an envelope of gas that keeps the planet warm and provides oxygen for breathing and carbon dioxide for photosynthesis. Finally, there is the fifth sys-tem, which contains huge quantities of ice at the poles and elsewhere, constituting the cryosphere. All five of these enormous and complex systems interact with one another to maintain the Earth as we know it. When observed from space, one of Earth's most obvious features is its abundant water. Although liquid water is present around the globe, the vast majority of the water on Earth, a whopping 96.5 percent, is saline (salty) and is not water humans, and most other animals, can drink without processing.

All of the liquid water on Earth, both fresh and salt, makes up the hydrosphere, but it is also part of other spheres. For instance, water vapor in the atmosphere is also considered to be part of the hydrosphere. Ice, being frozen water, is part of the hydrosphere, but it is given its own name, the cryosphere. Not only do the Earth systems overlap, they are also interconnected; what affects one can affect another. When a parcel of air in the atmosphere becomes saturated with water, precipitation, such as rain or snow, can fall to Earth's surface. That precipitation connects the hydrosphere with the geosphere by promoting erosion and weathering, surface processes that slowly break down large rocks into smaller ones. Over time, erosion and weathering change large pieces of rocks—or even mountains—into sediments, like sand or mud. The cryosphere can also be involved in erosion, as large glaciers scour bits of rock from the bedrock beneath them. Both the geosphere and hydrosphere provide the habitat for the biosphere, a global ecosystem that encompasses all the living things on Earth. The biosphere refers to the relatively small part of Earth's environment in which living things can survive.

It contains a wide range of organisms, including fungi, plants, and animals, that live together as a community. Biologists and ecologists refer to this variety of life as biodiversity. All the living things in an environment are called its biotic factors. The biosphere also includes abiotic factors, the nonliving things that organisms require to survive, such as water, air, and light.

The atmosphere—a mix of gases, mostly nitrogen and oxygen along with less abundant gases like water vapor, ozone, carbon dioxide, and argon—is also essential to life in the biosphere. Atmospheric gases work together to keep the global temperatures within livable limits, shield the surface of Earth from harmful ultraviolet radiation from the sun, and allow living things to thrive. It is clear that all of Earth's systems are deeply intertwined, but sometimes this connection can lead to harmful, yet unintended, consequences. One specific example of interaction between all the spheres is human fossil fuel consumption. Deposits of these fuels formed millions of years ago, when plants and animals—all part of the biosphere—died and decayed.

At that point, their remains were compressed within Earth to form coal, oil, and natural gas, thus becoming part of the geosphere. Now, humans—members of the biosphere—burn these materials as fuel to release the energy they contain. The combustion byproducts, such as carbon dioxide, end up in the atmosphere. There, they contribute to global warming, changing and stressing the cryosphere, hydrosphere, and biosphere. The many interactions between Earth's systems are complex, and they are happening constantly, though their effects are not always obvious.

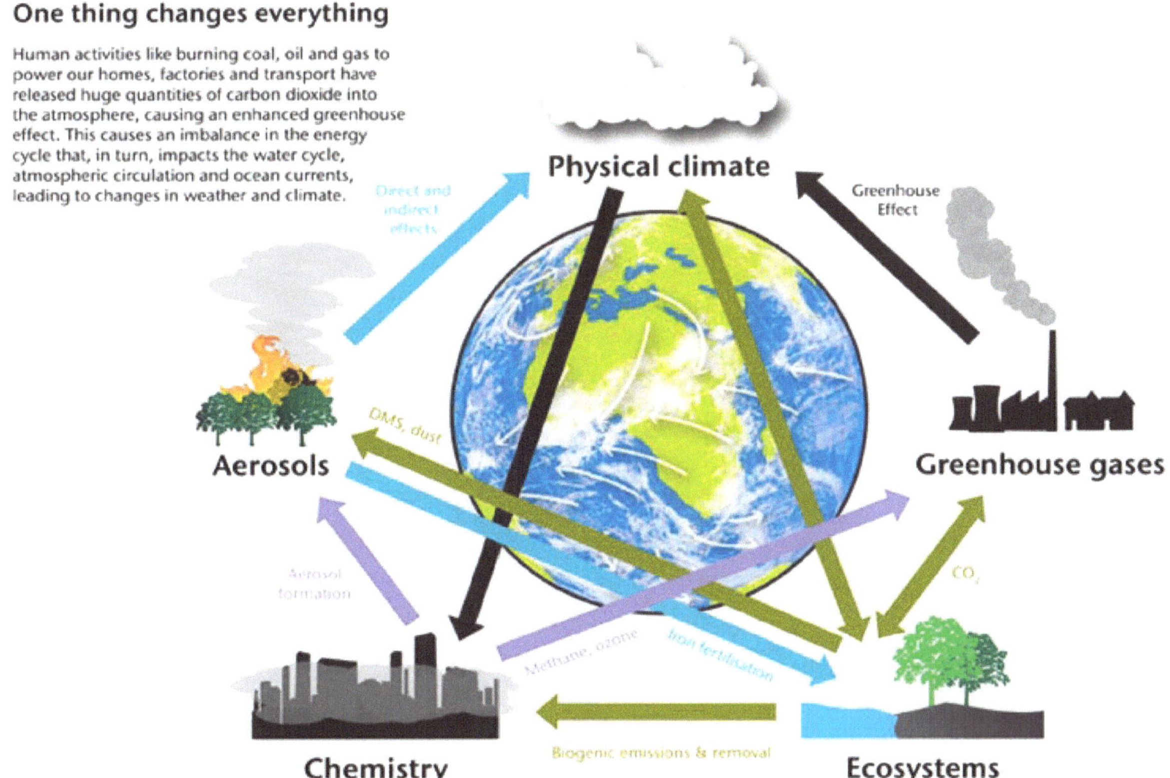

Reference link: https://education.nationalgeographic.org/resource/earths-systems/ &https://wmo.int/sites/default/files/styles/featured_image_x1_768x512/public/2023-11/Screenshot%202023-11-24%20at%2015.04.32.png?h=7b9e4ac6&itok=s7FHkE0v, https://blogs.reading.ac.uk/weather-and-climate-at-reading/files/2016/09/EarthSystem.png

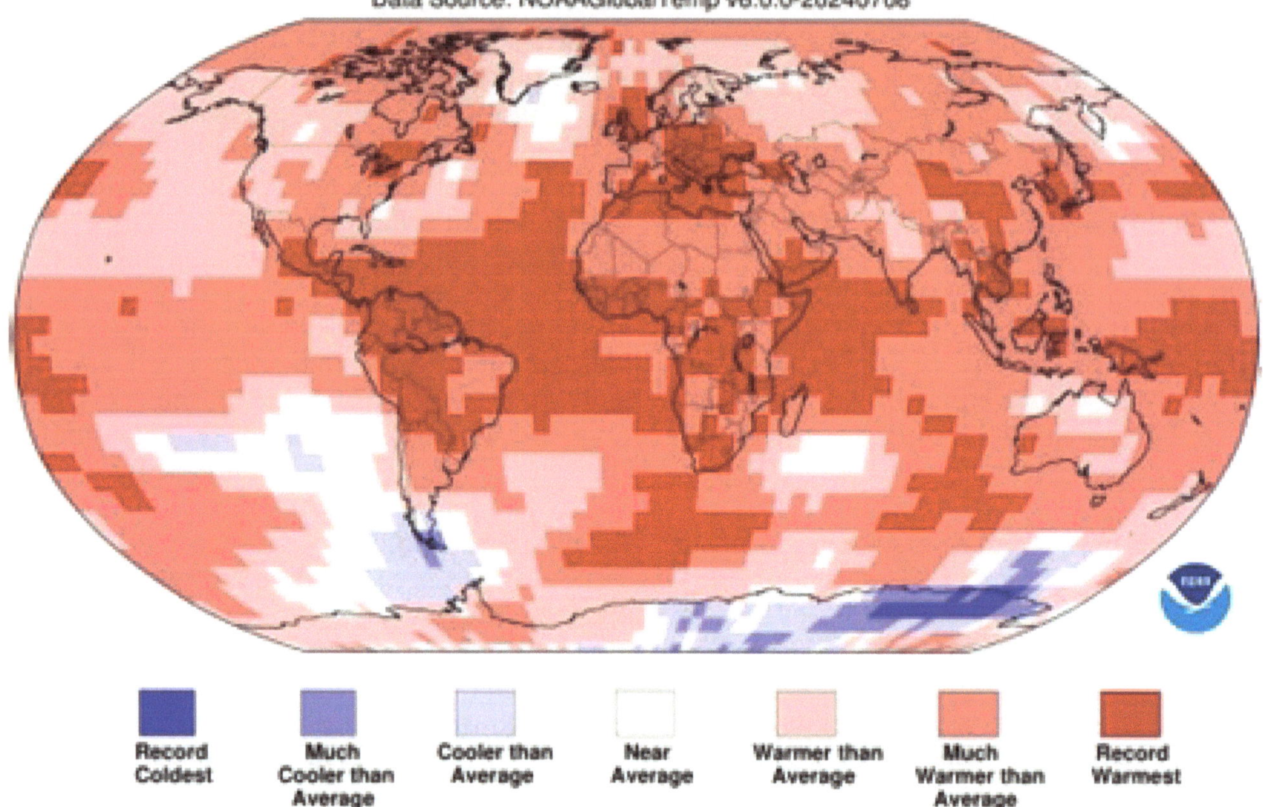

CLIMATOLOGY

Climatology is the study of the atmosphere's behavior and changes in temperature, pressure, and other atmospheric factors over time. Climatology is abranch of atmospheric science, but the study of climate can be related to everyotheraspectoftheEarth system including the geosphere (solidearth)and hydrosphere (terrestrial water reserves) because climate affects all of Earth'ssurface. Climatology is composed of two Greek words "Klima" and "Logos"." Klima"means inclination is latitude and"Logos"means science of study. Climatology is a science that describes and explains the nature of climate, why it differs from place to place,and how it relatesto other elements of the natural environment and human activities. It is the study of thevarieties of climates found on the Earth and their distribution over the surface of the Earth. Climatology is the study of the behavior of the atmosphere and changes in temperature, pressure, and other atmospheric factors overawhile.Climatology is a branch of atmospheric science, but the study of climate can be related to every other aspect of the Earth systemincluding the geosphere(solidEarth) and hydrosphere (terrestrial water reserves)because climate affects all of Earth's surface.

Climatology is categorized in to five basis subdivisions. Theyare:

Physical climatology:
This study focuse sonsolar energy and the exchange of energy-from one phase to the other.

Dynamic climatology:
This study focuses on the increase of green house gases as well as the sea surface temperature.

Applied climatology:
This study deals with theexplorationof the relationship of climatology to other fields of science.

Regional climatology:
It deals with the study of various climatic characteristics and their effect on the life, health, and economic condition of the people.

Synoptic climatology:
This study deals with the hemispheric climatic conditionsand the circulation patterns affecting the climate variation.

For example:

Synoptic climatology
Studies show upper-level wave patterns lead to heatwaves in central Europe.

Building climatology
Calculates the specifications of a drainage system based on the average rainfall in an area

Paleoclimate
Estimates temperature and precipitation conditions in Antarctica 100,000 years ago by analyzing the composition of an ice sheet layer.

Reference link: https://www.researchgate.net/publication/375743935_Definition_of_Cclimatology#:~:text=Definition%3A%20Climatology%20is%20the%20scientific,over%20long%20periods%20of%20time

ECOLOGY

Ecology is the study of the relationships between living organisms, including humans, and their physical environment. It seeks to understand the vital connections between plants and animals and the world around them. Ecology also provides information about the benefits of ecosystems and how we can use Earth's resources in ways that leave the environment healthy for future generations. Ecology first began gaining popularity in the 1960s when environmental issues were rising to the forefront of public awareness. Although scientists have been studying the natural world for centuries, ecology in the modern sense has only been around since the 19th century. Around this time, European and American scientists began studying how plants functioned and their effects on the habitats around them.

Eventually, this led to the study of how animals interact with plants and other animals and shape the ecosystems in which they live. Today, modern ecologists build on the data collected by their predecessors and continue to pass on information about ecosystems around the world. The information they gather continues to affect the future of our planet.Human activity plays an important role in the health of ecosystems all around the world. Pollution emitted from fossil fuels or factories can contaminate the food supply for a species, potentially changing an entire food web. Introducing a new species from another part of the world into an unfamiliar environment can have unintended and negative impacts on local life forms. These kinds of organisms are called invasive species.

Invasive species can be any form of living organism that is brought by humans to a new part of the world where they have no natural predators. The addition or subtraction of a single species from an ecosystem can create a domino effect on many others, whether that be from the spread of disease or overhunting. For example, some species have many weak feeding links, like omnivores, while others have fewer stronger links, like primary predators. Disrupting food webs can have a big impact on the ecology of a species or an entire ecosystem.

TYPES OF ECOLOGY

Aside from primary divisions, ecology has been divided into the following branches based on the level of organization, kind of environment or habitat, and taxonomic position. The different types of ecology are as follows:

Landscape Ecology – It deals with exchanging energy, living forms, materials, and other ecosystem components. It also portrays the impact and function of humans on a landscape.

Global Ecology – It studies the interactions between the Earth's habitats, including land, air, and sea.

Ecosystem Ecology – This is the integrated study of non-living and living components of ecological systems. It also covers their relationships within an ecosystem.

Habitat Ecology – It is concerned with studying various biosphere habitats, including forest ecology, grassland ecology, freshwater ecology, marine ecology, etc.

Organismal Ecology – The study of an organism's physiology, morphology, and conduct in relation to the environment.

Human Ecology – It specifically deals with the interaction between humans and their surroundings.

Community Ecology – It deals with interactions between the abiotic and biotic world. It also includes the vast diversity of interactions that happen between species.

Population Ecology – It deals with understanding, predicting, and explaining the rise and fall of a species as well as their distributions.

Behavioral Ecology – It explores the numerous ways in which animals evolve and respond to changes in their habitat.

Molecular Ecology – It applies genomics, molecular phylogenetics, molecular population genetics, and many more modern molecular concepts to traditional ecological concepts.

Source: https://education.nationalgeographic.org/resource/ecology/ Reference link: https://education.nationalgeographic.org/resource/ecology/

GEOLOGY

Geology describes the structure of the Earth on and beneath its surface and the processes that have shaped that structure. Geologists study the mineralogical composition of rocks to gain insight into their history of formation. Geology determines the relative ages of rocks found at a given location, while geochemistry (a branch of geology) determines their absolute ages. By combining various petrological, crystallographic, and paleontological tools, geologists can chronicle the geological history of the Earth as a whole. One aspect is to demonstrate the age of the Earth. Geology provides evidence for plate tectonics, the evolutionary history of life, and the Earth's past climates.

Solid fied lava flow in Hawaii

Geologists broadly study the properties and processes of Earth and other terrestrial planets. They use a wide variety of methods to understand the Earth's structure and evolution, including fieldwork, rock description, geophysical techniques, chemical analysis, physical experiments, and numerical modeling.]In practical terms, geology is important for mineral and hydrocarbon exploration and exploitation, evaluating water resources, understanding natural hazards, remediating environmental problems, and providing insights into past climate change. Geology is a major academic discipline, central to geological engineering, and plays an important role in geotechnical engineering. In the 1960s, it was discovered that the Earth's lithosphere, which includes the crust and the rigid uppermost portion of the upper mantle, is separated into tectonic plates that move across the plastically deforming, solid upper mantle, called the asthenosphere. This theory is supported by several types of observations, including seafloor spreading and the global distribution of mountain terrain and seismicity.

There is an intimate coupling between the movement of the plates on the surface and the convection of the mantle (that is, the heat transfer caused by the slow movement of ductile mantle rock). Thus, oceanic parts of plates and the adjoining mantle convection currents always move in the same direction because the oceanic lithosphere is the rigid upper thermal boundary layer of the convecting mantle. This coupling between rigid plates moving on the Earth's surface and the convecting mantle is called plate tectonics.

The development of plate tectonics has provided a physical basis for many observations of the solid Earth. Long linear regions of geological features are explained as plate boundaries:

Mid-ocean ridges are high regions on the seafloor where hydrothermal vents and volcanoes exist. These are seen as divergent boundaries, where two plates move apart. Arcs of volcanoes and earthquakes are theorized as convergent boundaries, where one plate subducts (moves under) another.

Transform boundaries, such as the San Andreas Fault system, are where plates slide horizontally past each other. Plate tectonics has provided a mechanism for Alfred Wegener's theory of continental drift, in which continents move across the Earth's surface over geological time. It also provided a driving force for crustal deformation and a new framework for observations in structural geology. The power of the theory of plate tectonics lies in its ability to combine all these observations into a single theory of how the lithosphere moves over the convecting mantle.

Advances in seismology, computer modeling, mineralogy, and crystallography at high temperatures and pressures provide insights into the internal composition and structure of the Earth.

Seismologists can use the arrival times of seismic waves to image the Earth's interior. Early advances in this field revealed the existence of a liquid outer core (where shear waves are unable to propagate) and a dense, solid inner core.

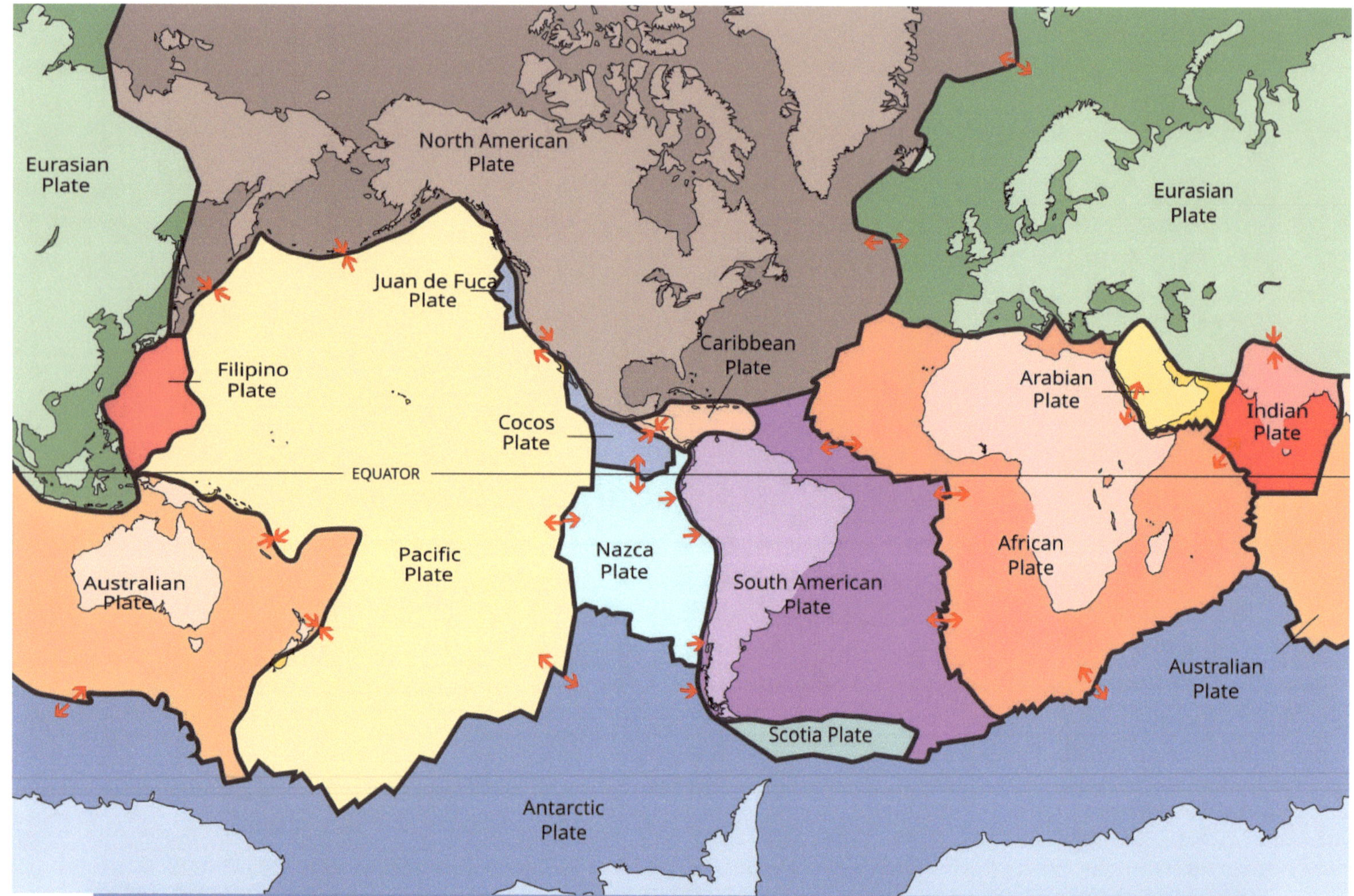

THE MAJOR TECTONIC PLATES OF THE EARTH

These advances led to the development of a layered model of the Earth, with the lithosphere (including the crust) on top, the mantle below (itself separated by seismic discontinuities at 410 and 660 kilometers), and the outer core and inner core below that.

More recently, seismologists have been able to create detailed images of wave speeds inside the Earth in the same way a doctor images a body in a CT scan. These images have provided a much more detailed view of the Earth's interior and have replaced the simplified layered model with a much more dynamic model.

Examples:
Environmental geology includes identifying and managing natural hazards such as earthquakes, floods, hillslope instability, soil erosion, subsidence, volcanoes, and wildfires.

OCEANOGRAPHY

Oceanography applies chemistry, geology, meteorology, biology, and other branches of science to the study of the ocean. It is especially important today as climate change, pollution, and other factors threaten the ocean and its marine life.

Oceanography is an interdisciplinary science where math, physics, chemistry, biology, and geology intersect. Traditionally, oceanography is discussed in terms of four separate but related branches:

Physical Oceanography – Involves the study of the properties (temperature, density, etc.) and movement (waves, currents, and tides) of seawater and the interaction between the ocean and the atmosphere.

Chemical Oceanography – Involves the study of the composition of seawater and the biogeochemical cycles that affect it.

Biological Oceanography – Involves the study of biological organisms in the ocean (including their life cycles and food production), such as bacteria, phytoplankton, and zooplankton. It also extends to the more traditional marine biology focus on fish and marine mammals.

Geological Oceanography – Focuses on the structure, features, and evolution of the ocean basins. Modeling allows oceanographers to examine the past and predict the future state of the ocean, such as circulation, air-sea interactions, sustainability of fisheries, and water quality. The knowledge gained from these measurements enables oceanographers to:

● Better predict changes in weather and climate.
● Improve forecasts for natural hazards (e.g., hurricanes) or man-made hazards (e.g., oil spills).
● Assess the impact of pollutants on water quality.
● Protect ocean water quality in the face of increasing human demands, such as fisheries, tourism, shipping, offshore oil and gas, and offshore wind farms.

For example: Chemical oceanographers study the composition of seawater and the chemical interactions of seawater with the atmosphere and the seafloor.

GLACIOLOGY

CONCEPT

Glaciology is the study of ice and its effects. Since ice can appear on or in the Earth, as well as in its seas, other bodies of water, and even its atmosphere, the purview of glaciologists is potentially very large. For the most part, however, glaciologists focus on great moving masses of ice called glaciers and the intervals of geologic history when glaciers and related ice masses covered relatively large areas of Earth. These intervals are known as ice ages, the most recent of which ended on the eve of human civilization's beginnings, just 11,000 years ago. The last ice age may not even be over, judging by the presence of large ice masses on Earth, including the vast ice sheet covering Antarctica. On the other hand, evidence gathered from the late twentieth century onward indicates the possibility of global warming brought about by human activity.

TYPES OF GLACIATIONS

Glaciologists, who study glaciers, have identified two types of glaciations that occur on the Earth's surface: alpine and continental.

Alpine Glaciation

Alpine glaciation refers to accumulations of ice confined to valleys. This ice moves down these valleys via mountainsides to reach the plains below. During their slow movement, alpine glaciers carve the landscape, deepening ravines and sharpening ridge lines. Consequently, the land near alpine glaciers appears rough and jagged.

HOW IT WORKS ICE

Ice is simply frozen water, but it is not as simple as it might appear. Glaciologists classify different types of ice based on their density, designating them with Roman numerals. The ice most familiar to us is classified as ice I. The ice in glaciers differs significantly from the ice in an ice cube or a winter pond. These differences arise from massive pressure, which reduces the air content of glacier ice. Glaciology is defined as the study of ice, its forms, and its effects. This gives glaciologists a much broader scope than geologists, meteorologists, or oceanographers, who are primarily concerned with the geosphere, atmosphere, and hydrosphere, respectively. Though ice is commonly associated with the hydrosphere, where it appears in oceans, rivers, and lakes, it is also found on and even beneath the solid Earth. In some situations, ice is even present in the atmosphere.

Continental Glaciation

Continental glaciation refers to large areas of ice coverage that accumulate over a wide range of landscapes, mainly in the Northern Hemisphere. Continental glaciers cover thousands of square miles and measure thousands of feet in thickness. Today, they are mainly found in Greenland and Antarctica. These glaciers tend to wear down the surrounding landscape into smooth surfaces.

Reference link https://www.worldatlas.com/articles/what-is-glaciology.html https://www.encyclopedia.com/earth-and-environment/ecology-andenvironmentalism/environmental-studies/glaciology
Source: https://www.worldatlas.com/articles/what-is-glaciology.html

SOLAR RADIATION

Solar radiation is the energy emitted by the Sun, which is sent in all directions through space as electromagnetic waves. Emitted by the surface of the Sun, this energy influences atmospheric and climatological processes. It is also directly and indirectly responsible for common phenomena, such as plant photosynthesis, keeping the planet at a temperature compatible with life, and wind formation, which is essential for generating wind power. The Sun emits energy in the form of short-wave radiation, which is weakened in the atmosphere by clouds and absorbed by gas molecules or suspended particles. After passing through the atmosphere, solar radiation reaches the oceanic and continental land surfaces, where it is either reflected or absorbed. Finally, the surface returns this energy to outer space in the form of long-wave radiation.

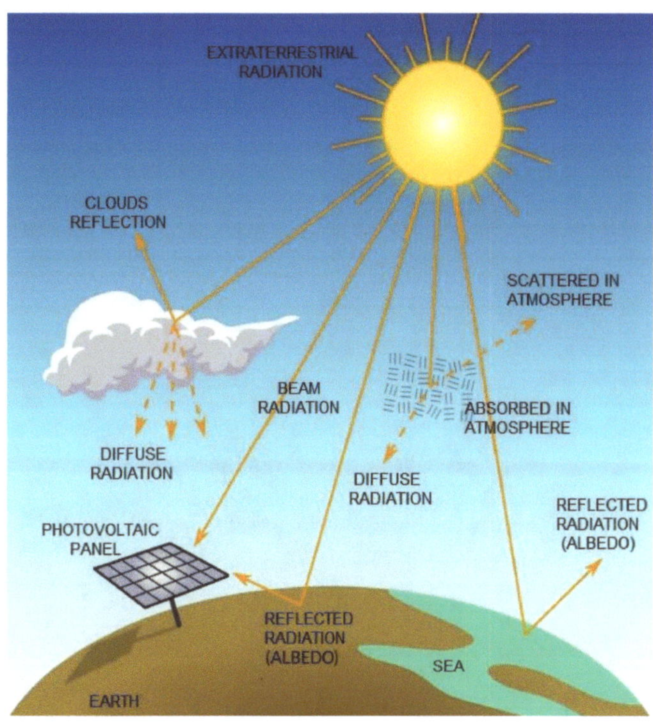

TYPES OF SOLAR RADIATION

Depending on the for min whichit reaches the Earth:

Direct solar radiation. This type of radiation penetrates the atmosphere and reaches the Earth'ssurface without dispersing at all on the way.

Diffuse solar radiation. This is the radiation that reaches the Earth'ssurface after having undergone multiple deviations in its trajectory, for example by gases intheatmosphere.

Reflected solar radiation. This is the fraction of solarradiationthatis reflected by the earth's surface itself, in a phenomenon known as the albedo effect.

Global normal irradiance (GNI) is the total irradiance from the Sun at thesurface of Earth at a given location with a surface element perpendicular to theSun.

Total solarirradiance(TSI)is a measure of the solarpower overall wavelength sperun it areaincident on the Earth'supper atmosphere. It is measured perpendicular to the incoming sunlight. The solar constant is a conventionalmeasureof meanTSIata distanceof oneastronomicalunit (AU).

Reference link: https://www.iberdrola.com/social-commitment/solar-radiation#:~:text=Solar%20radiation%20is%20the%20energy%20emitted%20by%20the%20Sun%2C%20which,influences%20atmospheric%20and%20climatological%20processes. Source: https://www.researchgate.net/figure/Representation-of-Solar-Radiation-Components_fig1_337408836

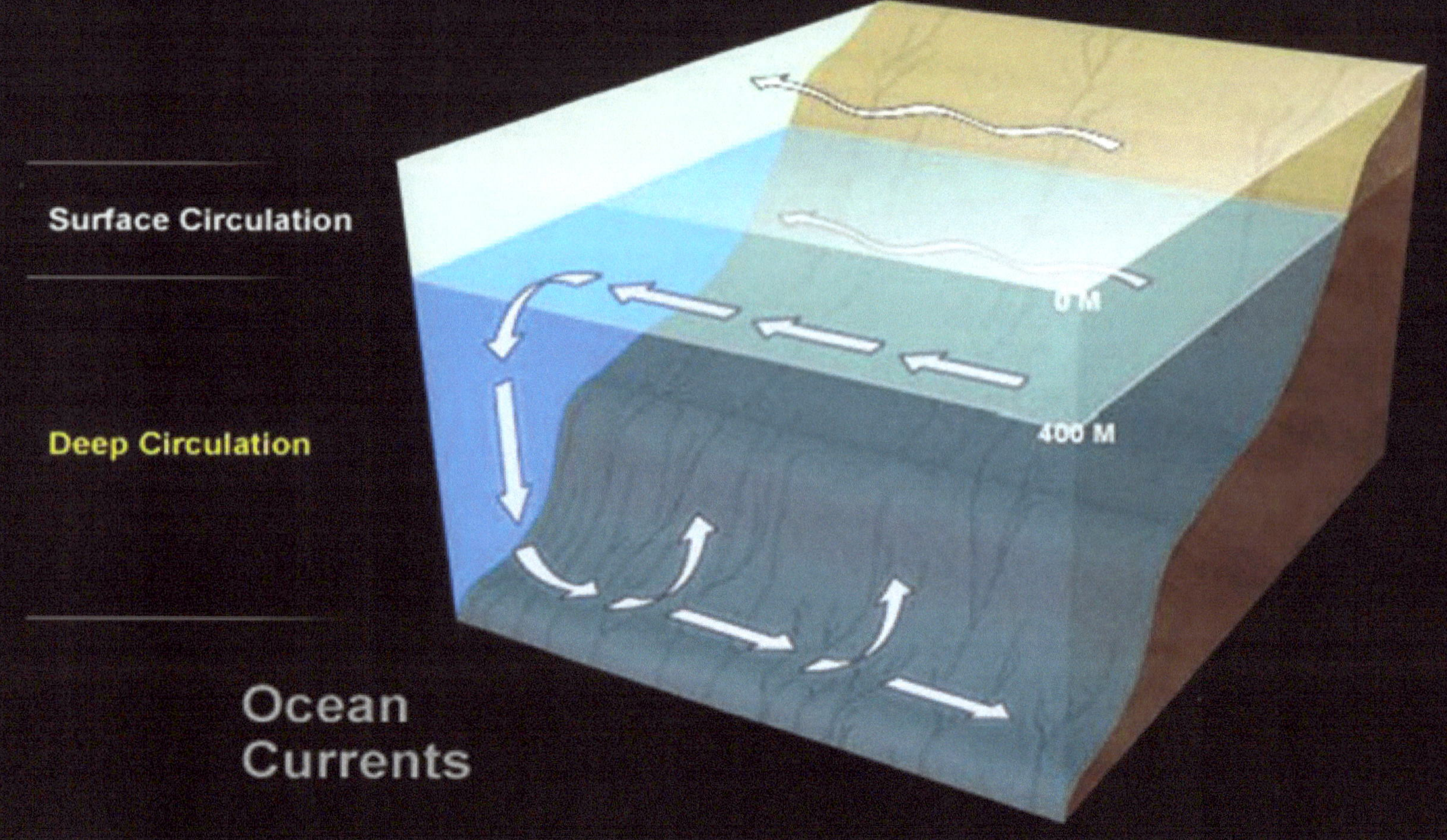

OCEAN CURRENTS

What are Ocean Currents?
Ocean currents are the continuous, predictable, directional movement of seawater. They represent massive flows of ocean water influenced by various forces and are similar to river flows in the oceans.

Ocean water moves in two directions:
Horizontally – Referred to as currents.
Vertically – Known as upwellings or downwellings.
Ocean currents significantly impact humankind and the biosphere due to their influence on climate.

Primary forces

Heating by solar energy: Heating by solar energy causes the water to expand. That is why, near the equator, the ocean water is about 8 cm higher in level than in the middlel atitudes.This causes as light gradient and watert ends to flow down the slope.

Wind: Wind blowing on the ocean's surface pushes the water to move.Friction between the wind and the water's surface affects the movement of the water body in it scourse.

Gravity: Gravity tends to pull the water down the pile and create gradient variation.

Differences in water density: It affects the vertical mobility of oceancurrents. Water with high salinity is denser than water with low salinity and in the same way, cold water is denser than warm water. Denser water tends to sink, while relatively lighter water tends to rise.

Types of Ocean Currents?

Classification of Ocean Currents Based on Depth

Surface Currents

Large-scale surface ocean currents are driven by global wind systems fueled by solar energy. These currents transfer heat from the tropics to the polar regions, influencing local and global climates. Surface currents constitute about 10% of all the water in the ocean and are confined to the upper 400 meters of the ocean.

Deepwater Currents

Differences in water density, caused by variations in water temperature (thermo) and salinity (haline), create deep ocean currents. This process is known as thermohaline circulation. Deepwater currents make up the remaining 90% of ocean water. These waters move around the ocean basins due to variations in density and gravity.

Deep waters sink into the deep ocean basins at high latitudes where temperatures are cold enough to increase density. This process initiates the global conveyor belt, a connected system of deep and surface currents that circulates around the globe over approximately 1,000 years.

Classification of Ocean Currents Based on Temperature

Cold Currents

Cold currents bring cold water into warm water areas. They are usually found on the west coasts of continents in low and middle latitudes (true for both hemispheres). In the Northern Hemisphere, they are also found on the east coasts of continents in higher latitudes.

Warm Currents

Warm currents bring warm water into cold water areas.They are typically observed on the east coasts of continents in low and middle latitudes (true for both hemispheres). In the Northern Hemisphere, they are also found on the west coasts of continents in higher latitudes.

Source: https://www.drishtiias.com/to-the-points/paper1/ocean-currents-1
Reference link: https://www.drishtiias.com/to-the-points/paper1/ocean-currents-1

KYOTO PROTOCOL

The Kyoto Protocol is based on the principles and provisions of the Convention and follows its annex-based structure. It binds only developed countries and places a heavier burden on them under the principle of "common but differentiated responsibility and respective capabilities", recognizing that they are largely responsible for the current high levels of greenhouse gas (GHG) emissions in the atmosphere.

In its Annex B, the Kyoto Protocol sets binding emission reduction targets for 37 industrialized countries and economies in transition and the European Union. Overall, these targets amount to an average 5 percent emission reduction compared to 1990 levels over the five years 2008–2012 (the first commitment period).

In Doha, Qatar, on 8 December 2012, the Doha Amendment to the Kyoto Protocol was adopted for a second commitment period, starting in 2013 and lasting until 2020. As of 28 October 2020, 147 Parties had deposited their instruments of acceptance, surpassing the threshold of 144 instruments required for the amendment to enter into force. The amendment officially. entered into force on 31 December 2020.

The amendment includes:

New commitments were established for Annex I Parties to the Kyoto Protocol who agreed to take on commitments during the second commitment period from 1 January 2013 to 31 December 2020.

These included:
A revised list of greenhouse gases (GHGs) to be reported by Parties in the second commitment period. Amendments to several articles of the Kyoto Protocol that specifically referenced issues related to the first commitment period, which needed updating for the second commitment period.
On 21 December 2012, the amendment was circulated by the Secretary-General of the United Nations, acting in his capacity as Depositary, to all Parties to the Kyoto Protocol, as per Articles 20 and 21 of the Protocol.

During the first commitment period, 37 industrialized countries andeconomies in transition and the European Community committed to reduce GHG emissions to an average offivepercent against 1990 levels.

During the second commitment period, Parties committed to reduce GHG emissions by at least 18 percent below 1990 levels in the eight years from 2013 to 2020; however, the composition of Parties in the second commitment period is different from the first.

The Kyoto mechanisms

One important element of the Kyoto Protocol was the establishment of flexiblemarket mechanisms, which are based on the trade of emissions permits. Underthe Protocol, countries must meet their targets primarily through national measures.

However,the Protocol also offers the man additional means to me ettheir targets by wayofthree market-based mechanisms:

International Emissions Trading	**Clean Development Mechanism(CDM)**

Joint Implementation (JI)

Monitoringe missiontargets

• The Kyoto Protocol also established a rigorous monitoring, review, andverification system, as well as a compliance system to ensuretransparency andhold Partiestoaccount.

• Under the Protocol, countries' actual emissions have to be monitored andpreciserecordshavetobekeptofthetradescarried out.

• RegistrysystemstrackandrecordtransactionsbyPartiesunderthemechanisms.

• The UN Climate Change Secretariat, based in Bonn, Germany, keeps an international transaction log to verify that transactions are consistent with the rules of the Protocol.

• Reporting is done by Parties by submitting annual emission inventoriesandnational reports under the Protocol at regular intervals.

• A compliance system ensures that Parties are meeting their commitments and helps them to meet their commitments if they have problems doing so.

Source: https://unfccc.int/kyoto_protocol
Reference link: https://unfccc.int/kyoto_protocol

DIFFERENCE BETWEEN
PARIS AGREEMENT AND KYOTO PROTOCOL

PARIS AGREEMENT	KYOTO PROTOCOL
The Paris Agreement was signed in 2016	The Kyoto Protocol was established in 1997
The Paris Agreement required both developing and developed nations to reduce their greenhouse emissions	The Kyoto Protocol primarily targeted industrialized nations as they were considered the primary emitters of greenhouse gases. Developing nations were exempt from the Kyoto Protocol
The objective of the Paris Agreement was to prevent the average global temperature from rising more than 2 degrees Celsius above pre-industrial levels	The objective of the Kyoto Protocol was to reduce greenhouse gases to 5.2%, below pre-1990 levels
The Paris Agreement was focused on reducing all anthropogenic greenhouse gases	The Kyoto Protocol was aimed at 6 major greenhouse gases such as carbon dioxide, methane, sulfur hexafluoride, HFCs, PFCs, and nitrous oxide
The goals of the Paris Agreement are to be achieved between 2025 and 2030	The first phase of the Kyoto Protocol lasted until 2012

Reference link: https://byjus.com/free-ias-prep/difference-between-kyoto-protocol-and-paris-agreement/

NATIONALLY DETERMINED CONTRIBUTIONS

The nationally determined contributions (NDCs) are commitments that countries make to reduce their greenhouse gas emissions as part of climate change mitigation. These commitments include the necessary policies and measures for achieving the global targets set out in the Paris Agreement. The nationally determined contributions (NDCs) are commitments that countries make to reduce their greenhouse gas emissions as part of climate change mitigation. These commitments include the necessary policies and measures for achieving the global targets set out in the Paris Agreement. The Paris Agreement has a long-term temperature goal which is to keep the rise In global surface temperature to well below 2 °C (3.6 °F) above pre-industrial levels. The treaty also states that preferably the limit of the increase should only be 1.5 °C (2.7 °F). To achieve this temperature goal, greenhouse gas emissions should be reduced as soon as, and by as much as, possible. To stay below 1.5 °C of global warming, emissions need to be cut by roughly 50% by 2030. This figure takes into account each country's documented pledges or NDCs. Probability that countries achieve their Paris Agreement Goals according to their nationally determined contributions NDCs embody efforts by each country to reduce national emissions and adapt to the impacts of climate change.The Paris Agreement requires each of the 193 Parties to prepare, communicate and maintain NDCs outlining what they intend to achieve. NDCs must be updated every five years.Prior to the Paris Agreement in 2015, the NDCs were referred to as intended nationally determined contributions (INDCs) and were non-binding. The INDCs were initial, voluntary pledges made by countries, whereas the NDCs are more committed but also not legally binding.

Source:https://assets.weforum.org/editor/6pZiNnUNSQJRSHXbV_YXhtCHtfidovLAT-wNLMTEYg8.PNG
Reference link: https://www.google.com/url?sa=t&source=web&rct=j&opi=89978449&url=https://en.wikipedia.org/wiki/Nationally_determined_contribution%23:~:text%3DThe%2520nationally%2520determined%2520contributions%2520(NDCs,out%2520in%2520the%2520Paris%2520Agreement.&ved=2ahUKEwjM1fTJ2viHAxVkTGwGHeEgDj0QFnoECBkQBA&usg=AOv-Vaw3D_9zPt0wuYbhBJHTOaJQiSour

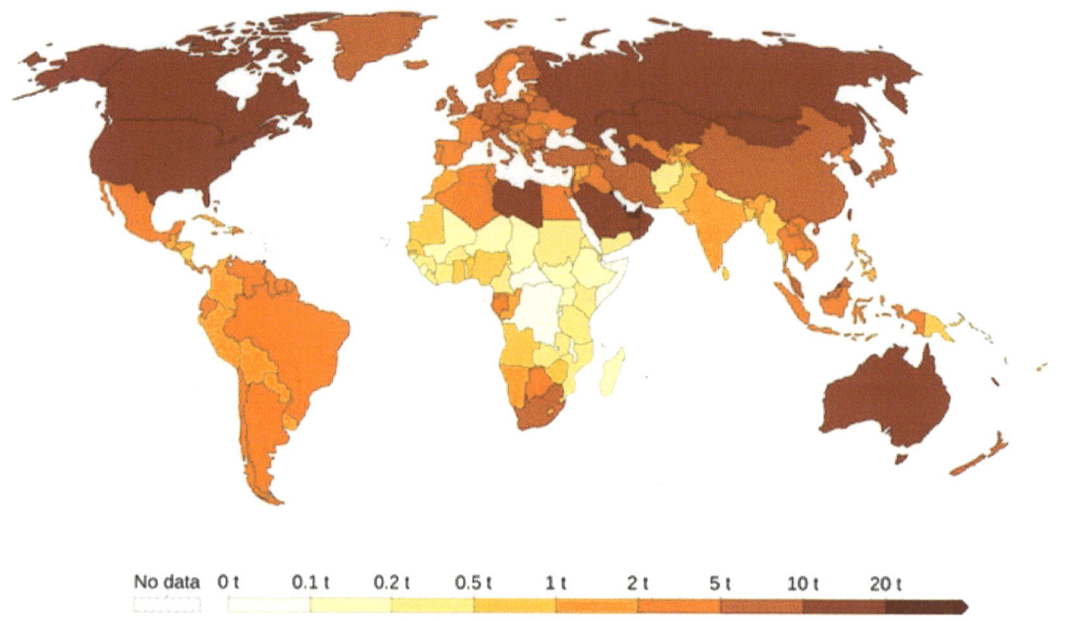

PER CAPITA EMISSIONS

Per capita emissions represent the emissions of an average person in a country or region – they are calculated as the total emissions divided by population. Emissions from international aviation and shipping are not included in any country or region's emissions.

Here's a breakdown of per capita emissions:
Calculated by dividing total GHG emissions by the total population
Usually measured in tons of carbon dioxide equivalent (tCO_2e) per person per year
Helps identify countries or regions with high emissions per person
Supports climate policy development and international agreements

Examples of per capita emissions (tCO_2e per person per year):
United States: around 16-17 tons
China: around 7-8 tons
European Union: around 6-7 tons
India: around 1.5-2 tons

Here are some additional facts about per capita emissions:

Variation: Per capita emissions vary greatly between countries, with some having much higher emissions per person than others.

Income: Generally, countries with higher incomes tend to have higher per capita emissions.

Population density: Countries with high population densities tend to have lower per capita emissions.

Energy consumption: Countries with high energy consumption per capita tend to have higher emissions.

Economic activity: Countries with strong industrial sectors or high levels of economic activity tend to have higher emissions.

Energy mix: Countries with a high reliance on fossil fuels tend to have higher emissions, while those with a higher share of renewables tend to have lower emissions.

Climate policy: Countries with strong climate policies and targets tend to have lower per capita emissions.

International cooperation: Global agreements and cooperation can help reduce per capita emissions by sharing best practices and technologies.

Reference link: https://ourworldindata.org/grapher/co-emissions-per-capita#:~:text=Per%20capita%20emissions%20represent%20the,total%20emissions%20divided%20by%20population.
Source: https://ourworldindata.org/grapher/co-emissions-per-capita#:~:text=Per%20capita%20emissions%20represent%20the,total%20emis-sions%20divided%20by%20population.

CONFERENCE OF PARTIES

The Conference of the Parties (COP) was established by the Convention as its main decision-making body. It is made up of governments and organizations such as the European Union and is responsible for guiding the Convention so that it can respond to global challenges and national needs.

AIMS

Reduce emissions

Scale up finance and support

Strengthen adaptation and resilience to climate impacts

History of COPs

Since 1995, representatives of **197 parties** (196 countries and the European Union) have been meeting to discuss the **climate issue**.

1992 — Rio de Janeiro Conference

1995 — Berlin, the first COP.

1997 — Kyoto, COP3. Kyoto Protocol.

2007 — "Bali, Action Plan"

2009 — COP15 in Copenhagen.

2010 — COP16 in Cancun. The Green Climate Fund.

2012 — COP18 in Doha. Ratified the Kyoto Protocol.

2013 — COP19 in Warsaw.

2015 — COP21 in Paris. An surpassed agreement.

2015 — The Paris Agreement entered into force.

2019 — COP 25 in Madrid.

2022 — COP27 in Sharm El Sheikh.

2023 — COP 28 in Dubai. COP26 in Glasgow.

Who participates in COP?

There a lot of people, its very busy! One edition can bring together up to 40,000 people. The main actors are the national delegations: the teams of negotiators that each country sends. They are the ones who will lead the debates, and upon whom the summits' final results depend.But they are not alone.A whole ecosystem is put in place during COPs, with a whole host of actors, exchanging with each other, defending their interests (or the greater good) and attempting to weigh in on the outcome of discussions.

They include: NGOs; companies, unions, representatives of indigenous people, journalists, scientists, etc. And finally, citizensyou and I, can in actual fact attend COPs, where areas dedicated to the general public provide spaces for exhibitions, debates and workshops centred around climate change. These public spaces are managed by host countries, while the "pro" areas are administered by the UN.

Objectives of COP

The great challenge of the climate COPs is to agree on the reduction of greenhouse gas (GHG) emissions due to human activities and on the measures to be taken to limit global warming. Therefore, we are talking about negotiations based on numbers: how much CO_2 we are releasinginto the atmosphere and by how many degrees we allow the planet to warm up.

Reference link: https://sustainability.decathlon.com/what-is-the-purpose-of-cop-27-and-all-the-others
https://www.undp.org/iran/conference-parties-cop
Source: https://laudatosimovement.org/news/an-overview-of-the-latest-cops-on-the-road-to-dubai-2023/

COP PRESIDENCY

The COP Presidency is the leadership position held by the country hosting the Conference of the Parties (COP) of the United Nations Framework Convention on Climate Change (UNFCCC). The COP President is responsible for:

1.Setting the tone:
Shaping the agenda and tone for the COP negotiations.
2. Facilitating negotiations:
Chairing and facilitating negotiations among countries.
3. Building consensus:
Working to build consensus and agreement among countries.
4. Representing the COP:
Serving as the official representative of the COP.
5. Coordinating logistics:
 Overseeing the organization and logistics of the COP conference.
6. Promoting implementation:
 Encouraging countries to implement their climate commitments.
7. Mobilizing support:
Mobilizing political will and support for climate action.
8. Engaging stakeholders:
Engaging with civil society, businesses, and other stakeholders.

The COP Presidency rotates among the five regional groups:
1. Africa
2. Asia
3. Central and Eastern Europe
4. Latin America and the Caribbean
5. Western Europe and Others

The COP President plays a crucial role in driving global climate action and ensuring the success of the COP conference.

COP presidency Goals:
*   Improve access to quality education for marginalized and disadvantaged groups
*   Enhance education outcomes and achievements
*   Increase investment in education and resource mobilization
*   Strengthen education systems and institutions
*   Promote global citizenship and sustainable development through education.

Reference link: https://unfccc.int/process-and-meetings/conferences/the-big-picture/what-are-united-nations-climate-change-conferences/how-cops-are-organized-questions-and-answers#:~:text=How%20is%20the%20COP%20President,formally%20to%20the%20UNFCCC%20secretariat.

UNITED NATIONS FRAMEWORK
CONVENTION ON CLIMATE CHANGE

The United Nations Framework Convention on Climate Change (UNFCCC) was adopted in 1992 with the ultimate aim of preventing dangerous human interference with the climate system. The 1997 Kyoto Protocol and 2015 Paris Agreement build on the Convention.

The United Nations Framework Convention on Climate Change (UNFCCC) is the UN process for negotiating an agreement to limit dangerous climate change. It is an international treaty among countries to combat "dangerous human interference with the climate system". The main way to do this is limiting the increase in greenhouse gases in the atmosphere. It was signed in 1992 by 154 states at the United Nations Conference on Environment and Development (UNCED), informally known as the Earth Summit, held in Rio de Janeiro. The treaty entered into force on 21 March 1994. "UNFCCC" is also the name of the Secretariat charged with supporting the operation of the convention, with offices on the UN Campus in Bonn, Germany.

The Paris Agreement was a considerable achievement for the international community. For the first time, a climate change agreement brought all countries into an ambitious undertaking to combat climate change by limiting global temperature rise to well below 2 degrees Celsius, and to strive for 1.5 degrees Celsius.

Every year, Parties to the Convention meet in Conference of the Parties (COPs), as well as in technical meetings throughout the year, to advance the aims and ambitions of the Paris Agreement and achieve progress in its implementation. UN Women engages in COPs and in subsidiary body meetings through:

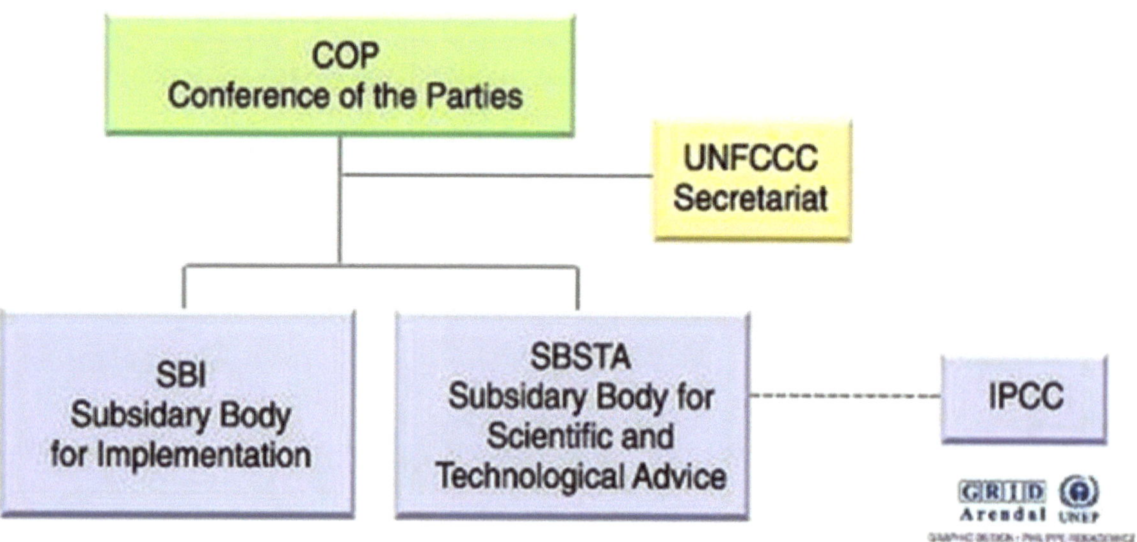

Preparing technical documents such as submissions, and analysis of COP documents from a gender perspective;

Providing technical advice and support to Parties on-site;

Bringing attention to gender issues through organizing or co-organizing side-events and exhibits and participating in panels and other events; and

Supporting the voices of women and girls from developing countries through facilitating their attendance to these meetings.

UNFCCC has several objectives:

1.   Stabilization of greenhouse gas concentrations: To stabilize atmospheric greenhouse gas concentrations at a level that prevents dangerous anthropogenic interference with the climate system.

2.   Prevention of dangerous climate change: To prevent dangerous climate change by reducing greenhouse gas emissions and enhancing sinks.

3.   Adaptation to climate change: To support countries in adapting to the impacts of climate change.

4.   Sustainable development: To promote sustainable development and eradicate poverty.

5.   Climate resilience and adaptation: To enhance climate resilience and adaptation in developing countries.

6.   Technology transfer and cooperation: To promote technology transfer and cooperation to support climate action.

7.   Climate finance: To mobilize financial resources to support climate action in developing countries.

8.   Capacity building: To build capacity in developing countries to support climate action.

9. Transparency and accountability: To promote transparency and accountability in climate action.

10. Global cooperation: To promote global cooperation and coordination on climate change.

The UNFCCC aims to address the global nature of climate change by bringing countries together to share knowledge, expertise, and resources to address this common challenge.

Here are the benefits of the UNFCCC:

Environmental Benefits:

- Reduced greenhouse gas emissions
- Slowing global warming
- Protection of ecosystems and biodiversity
- Preservation of natural resources
- Improved air and water quality

Socio-Economic Benefits:

- Job creation in the renewable energy sector
- Stimulated economic growth through sustainable development
- Improved public health through reduced air pollution
- Enhanced energy security and reduced dependence on fossil fuels
- Protection of vulnerable communities from climate change impacts

International Cooperation Benefits:

- Global cooperation and coordination on climate change
- Shared knowledge and expertise
- Technology transfer and cooperation
- Climate finance mobilization
- Capacity building and support for developing countries

Reference link: https://www.unwomen.org/en/how-we-work/intergovernmental-support/climate-change-and-the-environment/united-nations-framework-convention-on-climate-change
https://en.m.wikipedia.org/wiki/United_Nations_Framework_Convention_on_Climate_Change

SPECIAL CLIMATE CHANGE FUND

The Special Climate Change Fund (SCCF) was created in 2001 to address the specific needs of developing countries under the UNFCCC to adapt to the impact of climate change and increase resilience. It covers the incremental costs of interventions to address climate change adaptation relative to a development baseline.Adaptation to climate change is the top priority of the SCCF, although it can also support technology transfer and its associated capacity building activities. The SCCF is intended to catalyse and leverage additional finance from bilateral and multilateral sources, and is administered as a specialised trust fund by the Global Environment Facility (GEF).

The three strategic objectives for the SCCF are:

Reduce vulnerability and increase resilience through innovation and technology transfer for climate change adaptation

Mainstream climate change adaptation and resilience for systematic impact

Foster enabling conditions for effective and integrated climate change adaptation

What they do?

The Special Climate Change Fund, one of the world's first multilateral climate adaptation finance instruments, was created at the 2001 Conference of the Parties (COP) to the United Nations Framework Convention on Climate Change (UNFCCC) to help vulnerable nations in addressing these negative impacts of climate change. The SCCF is managed by the GEF and operates in parallel with the Least Developed Countries Fund (LDCF). Both funds have a mandate to serve the Paris Agreement. The GEF's new climate change adaptation strategy for the 2022-2026 period will focus SCCF support in the following two priority areas:

Supporting the adaptation needs of Small Island Developing States (SIDS)

The Small Island Developing States of the Caribbean, African and Indian Ocean, and the Pacific, are among the world's most vulnerable countries, due to a range of climatic and non-climatic factors. Salt water intrusion is severely impacting availability of drinking water as well as agricultural productivity on many islands. Sea level rise will worsen this situation, especially on low-lying islands, and, together with increased heavy rainfall, worsen damage from tropical storms to coastal infrastructure, settlements, and coastal ecosystems. Compounding these impacts, solutions are often difficult owing to SIDS' geographic isolation and limited land area. The Intergovernmental Panel on Climate Change Sixth Assessment Report refers to a range of projected adverse climate change impacts for SIDS, which will translate into direct adverse impacts on human security, health, infrastructure, ecosystems, agriculture and food, and the economy and livelihoods.

Some areas where the SCCF could offer adaptation support to SIDS include: storm and flood early warning systems; improved regional forecasts; nature-based solutions such as mangroves and other protective measures; enhanced resilience of roads, public infrastructure, and freshwater sources; climate-resilient aquaculture, fisheries, and diversified incomes; systemic resilience interventions in the food, urban and tourism space; climate resilient health (vector- and water-borne disease); and measures to build resilience, reduce fragility, and diversify the local economy, reducing dependence on imports; as well as mainstream climate resilience in policies and development planning; and build domestic capacity for adaptation.

Reference link: https://www.thegef.org/what-we-do/topics/special-climate-change-fund-sccf
https://unfccc.int/topics/climate-finance/resources/reports-of-the-special-climate-change-fund

SANTIAGO NETWORK

The Santiago network, established in December 2019 at COP 25, plays a crucial role in addressing climate change impacts in developing countries. It focuses on catalysing technical assistance from various organizations, bodies, networks, and experts to support developing countries in averting, minimizing, and addressing loss and damage caused by climate change.

Functions of the Santiago network

a)   Contributing to the effective implementation of the functions of the Warsaw International Mechanism, in line with the provisions in paragraph 7 of decision 2/CP.19 and Article 8 of the Paris Agreement, by catalysing the technical assistance of organizations, bodies, networks and experts;

(b) Catalysing demand-driven technical assistance including of relevant organizations, bodies, networks and experts, for the implementation of relevant approaches to averting, minimizing and addressing loss and damage in developing countries that are particularly vulnerable to the adverse effects of climate change by assisting in:

(i) Identifying, prioritizing and communicating technical assistance needs and priorities;

(ii) Identifying types of relevant technical assistance;

(iii) Actively connecting those seeking technical assistance with best suited organizations, bodies, networks and experts;

(iv) Accessing technical assistance available including from such organizations, bodies, networks and experts;

© Facilitating the consideration of a wide range of topics relevant to averting, minimizing and addressing loss and damage approaches, including but not limited to current and future impacts, priorities, and actions related to averting, minimizing, and addressing loss and damage pursuant to decisions 3/CP.18, and 2/CP.19, the areas referred to in Article 8, paragraph 4, of the Paris Agreement and the strategic workstreams of the five-year rolling workplan of the Executive Committee;

(d) Facilitating and catalysing collaboration, coordination, coherence and synergies to accelerate action by organizations, bodies, networks and experts, across communities of practices, and for them to deliver effective and efficient technical assistance to developing countries;

(e)Facilitating the development, provision and dissemination of, and access to, knowledge and information on averting, minimizing and addressing loss and damage, including comprehensive risk management approaches, at the regional, national and local level;

(f) Facilitating, through catalysing technical assistance, of organizations, bodies, networks and experts, access to action and support (finance, technology and capacity building) under and outside the Convention and the Paris Agreement, relevant to averting, minimising and addressing loss and damage associated with the adverse effects of climate change, including urgent and timely responses to the impacts of climate change.

Time line of Santiago network

December 2019:COP 25 in Madrid, Spain

Parties established the Santiago network as part of the WIM, to catalyse technical assistance of relevant organizations for the implementation of relevant approaches in developing countries that are particularly vulnerable to the adverse impacts of climate change (see Decision 2/CMA.2, para 43).

November 2021: COP 26 in Glasgow, UK
COP 26/CMA 3 decided on the functions of the Santiago Network and issued a call for submissions

November 2022: COP 27 in Sharm El-Sheikh, Egypt
COP 27/CMA 4 adopted the terms of reference for the Santiago network and established the Advisory Board of the Santiago network, with SB 58 to recommend a draft host agreement with the proposer, with a view to recommending it for consideration and approval by the governing body or bodies at their session(s) to be held in November 2023.

Vision

The vision of the Santiago Network is to catalyze the technical assistance of relevant organizations, bodies, networks and experts, for the implementation of relevant approaches for averting, minimize and addressing L&D at the local, national and regional level, in developing countries that are particularly vulnerable to the adverse effects of climate change (Decision 2/CMA.2, para 43).

Focus areas:

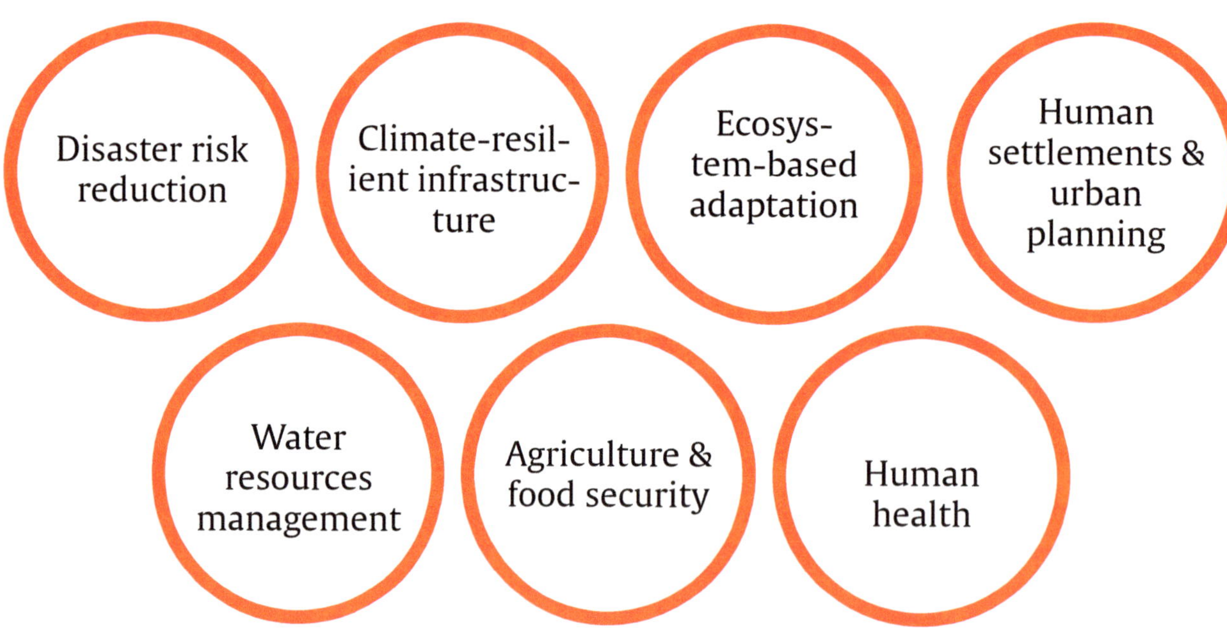

Source: https://www.lossanddamagecollaboration.org/pages/how-will-delayed-nominations-to-the-advisory-board-of-the-santiago-network-affect-full-operationalisation
Reference link: https://unfccc.int/santiago-network/about https://www.undrr.org/what-we-do/santiago-network

Content Editors

Thaiyalnayaki K
Second year M.com,
Department of Commerce,
Central university of Tamil Nadu.

Abinaya K
Second year M.com,
Department of Commerce,
Central university of Tamil Nadu.

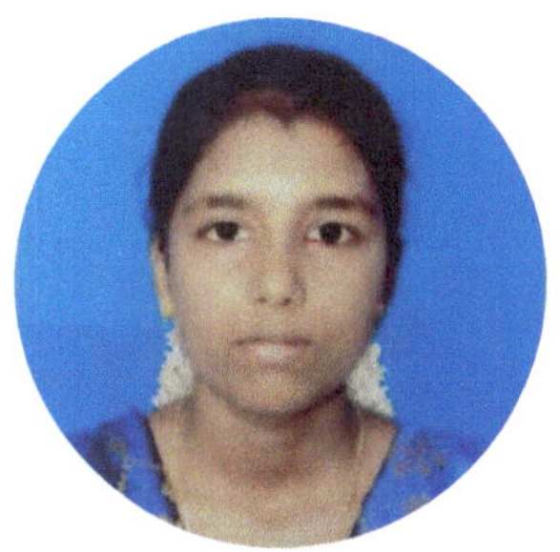

Keerthana I
Second year M.com,
Department of Commerce,
Central university of Tamil Nadu.

Layout Designers

Praveena E
Third year B.Voc, Department of
 Digital Journalism and Multimedia
Appications. Community College,
Central university of Tamil Nadu.

Jansi S
Third year B.Voc, Department of
 Digital Journalism and Multimedia
Appications. Community College,
Central university of Tamil Nadu.

Layout Artist

Naveen V
Faculty, Department of Digital Journalism and
Multimedia Appications. Community College,
Central university of Tamil Nadu.